TRADE ISSUES, POLICIES AND LAWS

REMANUFACTURED GOODS

U.S. ACTIVITIES, SECTOR STUDIES, AND GLOBAL MARKETS

Trade Issues, Policies and Laws

Additional books in this series can be found on Nova's website
under the Series tab.

Additional E-books in this series can be found on Nova's website
under the E-book tab.

Manufacturing Technology Research

Additional books in this series can be found on Nova's website
under the Series tab.

Additional E-books in this series can be found on Nova's website
under the E-book tab.

TRADE ISSUES, POLICIES AND LAWS

REMANUFACTURED GOODS

U.S. ACTIVITIES, SECTOR STUDIES, AND GLOBAL MARKETS

LORI A. SIMMONS
EDITOR

nova publishers
New York

For permission to use material from this book please contact us:
Telephone 631-231-7269; Fax 631-231-8175
Web Site: http://www.novapublishers.com

Library of Congress Cataloging-in-Publication Data

ISBN: 978-1-62618-501-2

Published by Nova Science Publishers, Inc. † New York

CONTENTS

PREFACE

The United States is the world's largest producer, consumer, and exporter of remanufactured goods. Remanufacturing is an industrial process that restores end-of-life goods to original working ("like new") condition. Remanufacturing occurs across a diverse range of industry sectors in the United States, but is more common in sectors making capital-intensive, durable products that have relatively longer product life cycles. The sectors that account for the majority of remanufacturing activity in the U.S. include aerospace, consumer products, electrical apparatus, heavy-duty and off-road (HDOR) equipment, information technology (IT) products, locomotives, machinery, medical devices, motor vehicle parts, office furniture, restaurant equipment, and retreaded tires. This book provides an overview of remanufactured goods activities, sector studies and global markets.

Chapter 1 - Remanufacturing is an industrial process that restores end-of-life goods to their original working condition. This report provides an overview of U.S. remanufactured goods industries and markets, estimates U.S. trade in remanufactured goods, and examines major factors that affected U.S. sales, trade, and investment during 2009–11. The United States is the largest remanufacturer in the world, and between 2009 and 2011, the value of U.S. remanufactured production grew by 15 percent to at least $43.0 billion, supporting 180,000 full-time U.S. jobs.

The remanufacturing-intensive sectors that account for the majority of remanufacturing activity in the United States include aerospace, consumer products, electrical apparatus, heavy-duty and off-road equipment, information technology products, locomotives, machinery, medical devices, motor vehicle parts, office furniture, restaurant equipment, and retreaded tires. U.S. exports of remanufactured goods totaled $11.7 billion in 2011; almost 40 percent of the total went to free trade agreement partners.

Foreign remanufacturers that have invested in the United States account for about one-sixth of U.S. trade in remanufactured goods and cores (the used or discarded component that is remanufactured).

Although the United States and Europe currently account for the bulk of remanufacturing activities and associated trade, other countries are developing their own remanufacturing industries. In foreign markets, regulatory barriers, import bans, and the lack of a common definition of remanufactured goods limit trade in remanufactured goods and cores.

Chapter 2 - The Motor & Equipment Manufacturers Association (MEMA) represents over 700 companies that manufacture and remanufacture motor vehicle parts for use in the light- and heavy-duty original equipment and aftermarket industries. Motor vehicle parts manufacturers are the nation's largest manufacturing sector, directly employing over 685,000 people across the country. MEMA represents its members through four affiliate associations: Automotive Aftermarket Suppliers Association (AASA); Heavy Duty Manufacturers Association (HDMA); Motor & Equipment Remanufacturers Association (MERA); and Original Equipment Suppliers Association (OESA).

MEMA, on behalf of its affiliate association, MERA, is pleased to submit comments in response to the Federal Trade Commission's (FTC) request for public comments regarding its revision to the "Guides for the Rebuilt, Reconditioned and Other Used Automobile Parts Industry" (Guides). In short, the Guides are outdated and outmoded because they suggest that remanufactured automotive products and various used automotive products are largely equivalent, thereby ignoring the significant advancements this country's remanufacturing industry has made with regard to consumer quality and in the area of international trade since the Guides were first adopted in 1979. Therefore, MEMA and MERA urge the Commission to overhaul the Guides to reflect these advancements and to encourage consumers to choose remanufactured products, which are produced using energy efficient and environmentally sound manufacturing processes. Such action would be consistent with other ongoing federal efforts.

Chapter 3 - In fiscal year 2011, federal civilian agencies reported about $975 million in maintenance and repair costs for approximately 588,000 vehicles that the agencies owned. These vehicles can be maintained and repaired using new or remanufactured parts. While there is no standard definition of a remanufactured vehicle part, the Federal Acquisition Regulation defines remanufactured parts as factory rebuilt to original specifications. Remanufactured vehicle parts tend to be less expensive than comparable new parts.

The principal remanufactured products in the motor vehicle sector are engines, transmissions, starter motors, alternators, steering racks, and clutches, according to the U.S. International Trade Commission.

In: Remanufactured Goods
Editor: Lori A. Simmons

ISBN: 978-1-62618-501-2

Chapter 1

REMANUFACTURED GOODS: AN OVERVIEW OF THE U.S. AND GLOBAL INDUSTRIES, MARKETS, AND TRADE*

U.S. International Trade Commission

ABSTRACT

Remanufacturing is an industrial process that restores end-of-life goods to their original working condition. This report provides an overview of U.S. remanufactured goods industries and markets, estimates U.S. trade in remanufactured goods, and examines major factors that affected U.S. sales, trade, and investment during 2009–11. The United States is the largest remanufacturer in the world, and between 2009 and 2011, the value of U.S. remanufactured production grew by 15 percent to at least $43.0 billion, supporting 180,000 full-time U.S. jobs.

The remanufacturing-intensive sectors that account for the majority of remanufacturing activity in the United States include aerospace, consumer products, electrical apparatus, heavy-duty and off-road equipment, information technology products, locomotives, machinery, medical devices, motor vehicle parts, office furniture, restaurant equipment, and retreaded tires. U.S. exports of remanufactured goods totaled $11.7 billion in 2011; almost 40 percent of the total went to free trade agreement partners.

Foreign remanufacturers that have invested in the United States account for about one-sixth of U.S. trade in remanufactured goods and cores (the used or discarded component that is remanufactured).

Although the United States and Europe currently account for the bulk of remanufacturing activities and associated trade, other countries are developing their own remanufacturing industries. In foreign markets, regulatory barriers, import bans, and the lack of a common definition of remanufactured goods limit trade in remanufactured goods and cores.

* This is an edited, reformatted and augmented version of U.S. International Trade Commission, USITC Publication 4356, dated October 2012.

ACRONYMS AND ABBREVIATIONS

AASA	Automotive Aftermarket Suppliers Association
AMDR	Association of Medical Device Reprocessors
APEC	Asia-Pacific Economic Cooperation
APRA	Automotive Parts Remanufacturers Association
ASEAN	Association of Southeast Asian Nations
ATS	Aviation Technical Services
CAFTA-DR	Dominican Republic-Central America-United States Free Trade Agreement
CBI	confidential business information
CDMA	code division multiple access
Census	U.S. Census Bureau
CFR	Code of Federal Regulations
Commission	U.S. International Trade Commission
CRTAI	Cartridge Recyclers and Traders Association of India
CT	computed tomography
DDR	Detroit Diesel Remanufacturing
DIY	do-it-yourself
EASA	European Aviation Safety Agency
EC	European Commission
EPZ	export processing zone
ETIRA	European Toner and Inkjet Remanufacturers Association
EU	European Union
FAA	Federal Aviation Administration
FDA	Food and Drug Administration
FDI	foreign direct investment
FTA	free trade agreement
FTZ	free trade zone
GE	General Electric
GSM	global system for mobile communications
HDMA	Heavy Duty Manufacturers Association
HDOR	heavy-duty and off-road
ICRRA	Indian Cartridge Remanufacturers and Recyclers Association
HS	Harmonized Commodity Description and Coding System
HTS	Harmonized Tariff Schedule of the United States
IITC	International Imaging Technology Council
IT	information technology
MDIC	Ministry of Development, Industry, and Foreign Trade (Brazil)
MEMA	Motor and Equipment Manufacturers Association
MEP	Ministry of Environmental Protection (China)
MERA	Motor and Equipment Remanufacturers Association
MIIT	Ministry of Industry and Information Technology (China)
MITA	Medical and Imaging Technology Alliance
MOFCOM	Ministry of Commerce (China)

MRO	maintenance and repair organization
MRI	magnetic resonance imaging
NAFTA	North American Free Trade Agreement
NAICS	North American Industry Classification System
NDRC	National Development and Reform Commission (China)
NEMA	National Electrical Manufacturers Association
NIST	National Institute of Standards and Technology
OE	original equipment
OEM	original equipment manufacturer
OES	original equipment supplier
OESA	Original Equipment Suppliers Association
OPI	OEM Product-Services Institute
OTR	off-the-road
PC	personal computer
PCRR	PC Rebuilders and Recyclers
PET	positron emission tomography
R&D	research and development
RIC	Remanufacturing Industries Council
RIT	Rochester Institute of Technology
RLA	Reverse Logistics Association
RoHS	Restriction of the Use of Certain Hazardous Substances in Electrical and Electr Equipment (Restrictions on the Use of Hazardous Substances)
RPK	revenue passenger kilometer
RSE	relative standard error
SMEs	small and medium-sized enterprises
SUD	single-use devices
TIA	Tire Industry Association
TPP	Trans-Pacific Partnership
TPR	third-party reprocessors
TRIB	Tire Retread & Repair Information Bureau
UK	United Kingdom
USDOC	U.S. Department of Commerce
USITC	U.S. International Trade Commission
USTR	U.S. Trade Representative
VAT	value-added tax
WEEE	Waste Electrical and Electronic Equipment
WHO	World Health Organization

GLOSSARY

CAFTA-DR: The Dominican Republic-Central America-United States Free Trade Agreement. Partners include the Dominican Republic, Costa Rica, El Salvador, Guatemala, Honduras, Nicaragua, and the United States.

Closed-loop supply chain: Refers to the management and supply of cores to be remanufactured. In this type of supply chain, remanufacturers source used or end-of-life manufactured products as the principal material to be remanufactured (restored to original working condition). Various techniques are used to manage the supply of cores, all based generally on the concept of a core deposit.

Core deposit: Generally, an additional charge incorporated into the price of a remanufactured good. Once a core (see below) is returned to the remanufacturer, the deposit is credited back to the customer that returned the core (also known as a transaction on an "exchange basis"). Such a system encourages core returns to ensure an adequate and reliable source of core supply.

Cores: Used goods that are the primary component input for remanufactured goods. Typically, these goods are at the end of their useful life.

Fleet customers: Organizations that purchase high volumes of motor vehicles for transportation purposes. Encompasses a broad range of groups, including logistics firms, car rental companies, retail chains with their own heavy trucking fleets, police departments, and government agencies.

Foreign direct investment (FDI): Investment funds used to purchase a 10 percent or greater ownership interest in another firm, and any subsequent funds used to increase this ownership share. Also includes any lent funds, as well as earnings owed that were reinvested in that firm. Does not include direct investment stocks made before the 2009–11 study period.

Foreign remanufacturer invested in the United States: A firm that reported (1) that one or more foreign entities had a substantial (10 percent or greater) ownership interest in it as of December 31, 2011; and (2) that at least one of these foreign entities had remanufacturing activities in countries other than the United States. An "entity" is a corporation, limited-liability company, partnership, sole proprietorship, or any subsidiary of any of these forms of businesses.

Full-time workers: The number of a firm's employees at facilities located in the United States who perform work related to all remanufacturing activity, on a full-time equivalent (FTE) basis. (FTE=the total number of annual regular hours worked by employees divided by the number of compensable hours applicable to each calendar year.) Employees can include production and related workers, warehouse/distribution workers, sales staff, managers, supervisors, technicians, and office workers.

Inbound FDI: Investment funds from a foreign entity used to purchase a 10 percent or greater ownership interest in a firm's U.S. remanufacturing activities, and any subsequent funds from the foreign entity used to increase this ownership share. Also includes any funds the foreign entity lent to that firm, as well as earnings owed to the foreign entity that were reinvested in that firm. Does not include direct investment stocks in U.S. remanufacturing activities made before the 2009–11 study period.

Independent or third-party remanufacturer: A firm that remanufacturers a good that was originally produced by another firm.

Investment: Capital expenditures for plant construction, improvements to existing plant and equipment, and purchases of new or existing plant, property, machinery, and equipment. Also includes capital expenditures for property, plant, and equipment related to research and development.

Maintenance and repair organizations (MROs): Firms that overhaul engines, perform heavy maintenance on airframes, repair components, and carry out line maintenance on aircraft to restore aircraft to airworthy condition. (Line maintenance is performed at the gate or on the ramp between flights.)

Outbound FDI: Investment funds from a U.S.-based firm used to purchase a 10 percent or greater ownership interest in foreign remanufacturing activities, and any subsequent funds used to increase this ownership share. Also includes any funds the U.S.-based firm lent to the foreign entity, as well as earnings owed to the U.S-based firm that were reinvested in that foreign entity. Does not include direct investment stocks in foreign remanufacturing activities made before the 2009–11 study period.

Remanufactured goods: Non-agricultural goods that are entirely or partially comprised of parts that (i) have been obtained from the disassembly of used goods; and (ii) have been processed, cleaned, inspected, and tested to the extent necessary to ensure they have been restored to original working condition or better; and for which the remanufacturer has issued a warranty.

Remanufacturer: A firm that restores end-of-life goods to original working condition or better.

Remanufacturing activities: Activities related to (but not including) the production of remanufactured goods, including disassembling cores into parts; cleaning, inspecting, and testing parts; and conducting wholesale activities involving remanufactured goods or cores (see "Wholesalers").

Remanufacturing: An industrial process that restores end-of-life goods to original working condition or better. Firms that provide remanufacturing services to restore end-of-life goods to original working condition are considered producers of remanufactured goods.

Small and medium-sized enterprises (SMEs): Firms that employ 20–499 workers. The Commission did not systematically survey firms with fewer than 20 employees.

U.S. production of remanufactured goods: The value of shipments (including commercial shipments, internal consumption, and transfers to related firms) of remanufactured goods produced in the United States.

U.S. free trade agreement (FTA) partners: Includes the following countries that were U.S. FTA partners during 2009–11: Australia, Bahrain, Canada, Chile, Costa Rica, Dominican Republic, El Salvador, Guatemala, Honduras, Israel, Jordan, Mexico, Morocco, Nicaragua, Oman, Peru, and Singapore.

Wholesalers: Firms that sell, trade (import or export), warehouse, and/or distribute remanufactured goods or cores, but do not produce remanufactured goods. Also included are firms that purchase cores from a domestic firm.

EXECUTIVE SUMMARY

The United States is the world's largest producer, consumer, and exporter of remanufactured goods. Remanufacturing is an industrial process that restores end-of-life goods to original working ("like new") condition. Remanufacturing occurs across a diverse range of industry sectors in the United States, but is more common in sectors making capital-intensive, durable products that have relatively longer product life cycles. The sectors that account for the majority of remanufacturing activity in the United States include (in alphabetical order) aerospace, consumer products, electrical apparatus, heavy-duty and off-road (HDOR) equipment, information technology (IT) products, locomotives, machinery, medical devices, motor vehicle parts, office furniture, restaurant equipment, and retreaded tires.

Main Findings and Observations

U.S. production of remanufactured goods totaled at least $43.0 billion in 2011.

U.S. production of remanufactured goods grew from $37.3 billion in 2009 to $43.0 billion in 2011.[1] Remanufactured goods are estimated to have accounted for about 2 percent of total sales of all products (new and remanufactured) in the industry sectors noted above during the 2009–11 period. U.S. production of remanufactured aerospace products, HDOR equipment, and motor vehicle parts together accounted for 63 percent of total U.S. production of remanufactured goods (table ES.1). Small and medium-sized enterprises (SMEs) comprise an important share of remanufacturing production and trade. For instance, SMEs are estimated to have accounted for 25 percent ($11.1 billion) of U.S. production of remanufactured goods, and 17 percent ($1.8 billion) of U.S. exports in 2011.

Table ES.1. Remanufactured goods production, investment, employment, exports, imports, and remanufacturing intensity, 2011

Sector (ranked by production value)	Production (thousands $)	Investment (thousands $)	Employment (full-time workers)	Exports (thousand $)	Imports (thousand $)	Remanufacturing intensity (%)[a]
Aerospace	13,045,513	90,471	35,201	2,589,543	1,869,901	2.6
HDOR equipment	7,770,586	162,746	20,870	2,451,967	1,489,259	3.8
Motor vehicle parts	6,211,838	105,684	30,653	581,520	1,481,939	1.1
Machinery	5,795,105	711,008	26,843	1,348,734	268,256	1.0
IT products	2,681,603	17,503	15,442	260,032	2,756,475	0.4
Medical devices	1,463,313	31,260	4,117	488,008	110,705	0.5
Retreaded tires	1,399,088	23,874	4,880	18,545	11,446	2.9
Consumer products	659,175	4,948	7,613	21,151	360,264	0.1
All other[b]	3,973,923	67,537	22,999	224,627	40,683	1.3
Wholesalers	([c])	8,294	10,891	3,751,538	1,874,128	([c])
Total	43,000,144	1,223,326	179,509	11,735,665	10,263,056	2.0

Source: USITC staff calculations of weighted responses to the Commission questionnaire. Figures for U.S. exports and imports of retreaded tires are taken from official statistics of the U.S. Department of Commerce for HTS subheadings 4012.11, 4012.12, 4012.13, and 4012.19. Totals may not sum due to rounding.

[a]Total value of shipments of remanufactured goods as a share of total sales of all products within that sector. [b]Includes remanufactured electrical apparatus, locomotives, office furniture, and restaurant equipment. [c]Wholesalers do not produce remanufactured goods, but rather sell or trade (export and import) them.

U.S. remanufacturing activities supported at least 180,000 full-time jobs in the United States in 2011.

U.S. remanufacturing employment grew from about 166,000 in 2009 to 180,000 in 2011. U.S. production of remanufactured aerospace products, HDOR equipment, and motor vehicle parts together supported about 87,000 full-time jobs. SMEs are estimated to have accounted for 36 percent (65,500 workers) of U.S. remanufacturing employment in 2011.

U.S. exports of remanufactured goods totaled $11.7 billion in 2011, up 50 percent compared with 2009.

The United States is a net exporter of remanufactured goods. U.S. exports of remanufactured goods increased by over 50 percent from $7.5 billion in 2009 to $11.7 billion in 2011, while imports grew by 64 percent from $6.3 billion to $10.3 billion during the same period. Remanufactured aerospace products, HDOR equipment, and machinery accounted for the largest share of U.S. exports of remanufactured goods. Although remanufactured motor vehicle parts were among the largest remanufacturing sectors in terms of production, the sector's exports were relatively small because the sector tends to produce and sell mainly for the domestic market. Canada, the European Union (EU), and Mexico are the leading destinations for U.S. exports of remanufactured goods (table ES.2). Wholesalers (firms that do not produce remanufactured goods, but rather sell or trade them) are important players in remanufactured goods trade, accounting for nearly one-third of U.S. exports.

Table ES.2. Remanufactured goods trade: Leading export destinations and import sources for remanufactured oods and cores, 2011 (destinations and sources ranked by value

Sector (ranked by production value)	Leading destinations for U.S. exports of remanufactured goods	Leading FTA markets for U.S. exports of remanufactured goods	Leading suppliers of remanufactured goods to the United States
Aerospace	EU Canada Japan	Canada Singapore Mexico	EU Canada Japan
HDOR equipment	Canada Australia Mexico	Canada Australia Mexico	Mexico Canada EU
Motor vehicle parts	Canada Saudi Arabia Mexico	Canada Mexico Australia	Mexico EU Canada
IT products	EU Canada Hong Kong	Canada Mexico Singapore	Mexico China Canada
Medical devices	EU Canada Brazil	Canada Singapore Mexico	EU Mexico Canada
Retreaded tires	Mexico Canada Vietnam	Mexico CAFTA-DR[a] Canada	Canada EU Korea
Other remanufacturing sectors[b]	Mexico EU Canada	Mexico Canada CAFTA-DR	Switzerland China Mexico

Source: USITC staff calculations of weighted responses to the Commission questionnaire.

[a]Dominican Republic-Central America-United States Free Trade Agreement partners.

[b]Includes (in alphabetical order) the consumer products, electrical apparatus, locomotive, machinery, office furniture, and restaurant equipment remanufacturing sectors.

Almost 40 percent of U.S. exports of remanufactured goods went to free trade agreement (FTA) partners.

U.S. FTA partners are important markets for exports from every remanufacturing sector. U.S. exports of remanufactured goods to FTA countries grew by 50 percent from $3.0 billion in 2009 to $4.4 billion in 2011 (but not necessarily because of FTA provisions). Canada and Mexico, the United States' partners in the North American Free Trade Agreement, accounted for over 70 percent of U.S. exports of remanufactured goods to FTA partners in 2011 (figure ES.1). While the United States exported a variety of remanufactured goods to its FTA partners, goods in a few sectors tended to predominate. Remanufactured HDOR equipment, motor vehicle parts, and aerospace products comprise almost 50 percent of total U.S. exports of remanufactured goods to Canada. Pacific Rim countries Australia and Singapore are also important FTA markets for remanufactured HDOR equipment. Over two-thirds of U.S. exports of remanufactured goods to Australia are remanufactured HDOR equipment used for mining.

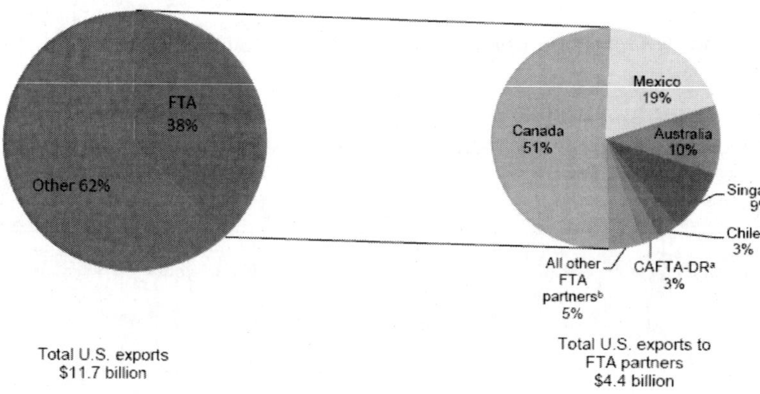

Source: USITC staff calculations of weighted responses to the Commission questionnaire.
[a]Dominican Republic-Central America-United States Free Trade Agreement partners.
[b]Bahrain, Israel, Jordan, Morocco, Oman, and Peru.

Figure ES.1. U.S. exports of remanufactured goods, 2011.

Foreign remanufacturers invested in the United States account for a notable portion of trade in remanufactured goods and cores.

Foreign remanufacturers that had invested in the United States accounted for about 17 percent of U.S. exports and 15 percent of U.S. imports of remanufactured goods during 2009–11. Remanufactured machinery accounted for over one-half of these exports in 2011; Mexico, the EU, and Canada were major markets. Remanufactured motor vehicle parts accounted for 43 percent of U.S. imports by foreign remanufacturers invested in the United States. The EU was a large supplier of these types of imports.

Trade in remanufactured goods is significantly affected by regulatory barriers in foreign markets, and by the price and availability of cores.

Widespread regulatory barriers in foreign markets strongly affect trade in both remanufactured goods and cores (the used goods to be remanufactured). A particular problem is the absence of a commonly accepted legal definition of remanufactured goods, leading many customs authorities to treat remanufactured goods as used products. Imports of used products are often more heavily restricted abroad than new ones; in many cases, they are prohibited. Similarly, cores are frequently classified as used products and are often banned from importation. Such bans force remanufacturers that produce remanufactured goods in foreign markets to rely on cores sourced from within those markets, where supplies are often limited.

Although numerous factors affect the competitiveness of U.S. remanufacturers, the availability and relative price of cores is critical to all sectors and in all markets.

The broad scope of remanufacturing activities reflects a diversity of industrial sectors, each of which responds to the particular combination of demand and supply factors in its own market segment (table ES.3). Domestic demand for remanufactured goods is driven by the comparative price of remanufactured goods versus new goods, the availability of lower-priced new alternatives, and customer perceptions about price and quality. For example, remanufacturers of short life-cycle consumer products like cell phones and DVD players are most concerned with the availability of competing low-cost, more technologically sophisticated new products. Factors that influence the supply of remanufactured goods vary between sectors. They include the availability, cost, and management of cores; manufacturing costs, including transportation and labor; the availability of skilled labor; and, in foreign markets, regulatory barriers. Core management is critical in all remanufacturing sectors (both domestic and foreign); firms in these sectors use a variety of techniques to manage their supply of inputs, based largely on the concept of a closed-loop supply chain. In this supply chain, remanufacturers source cores, which are the principal materials to be remanufactured (figure ES.2).

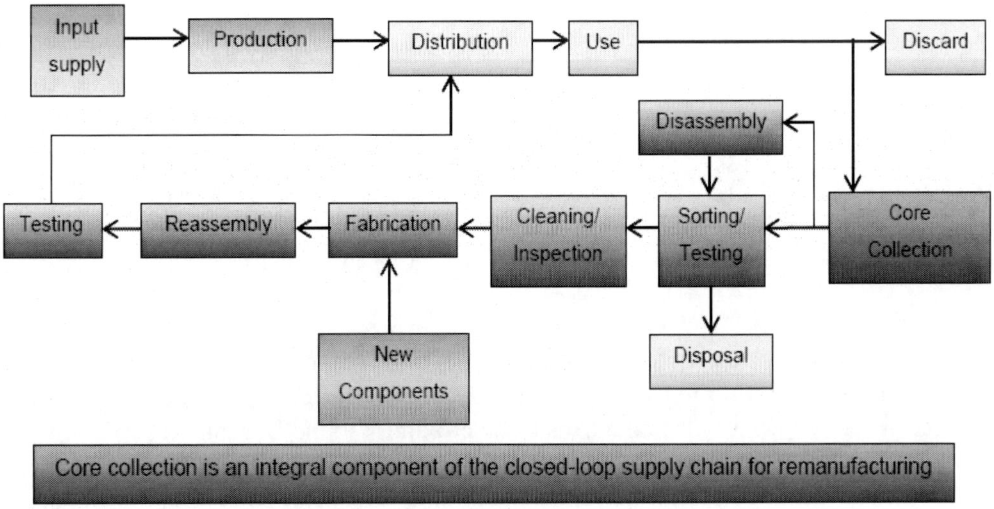

Source: Derived from Japke, Practice Guidelines, 2009, 10–11.

Figure ES.2. In contrast to manufacturing (blue), remanufacturing (green) relies on a closed-loop supply chain for cores.

Table ES.3. Most frequently cited "extremely important" factors that affect the ability of U.S. remanufacturers to compete in U.S. and foreign markets (top three factors ranked from most to least cited)

Sector (ranked by production value)	The U.S. market	Foreign markets
Aerospace	Availability of skilled workers Availability of cores Labor costs	Regulatory barriers in foreign markets Licensing or certification requirements Transportation costs
HDOR equipment	Transportation costs Environmental regulations Availability of cores	Foreign market tariffs Transportation costs Regulatory barriers in foreign markets
Motor vehicle parts	Labor costs Transportation costs Availability of cores	High price of cores Cost of compliance with products' environmental standards Licensing or certification requirements
Machinery	Availability of skilled workers Healthcare costs Unfavorable tax treatment	High price of cores Transportation costs Regulatory barriers in foreign markets
IT products	High price of cores Customer preferences for new products Availability of low-cost new products	Transportation costs Lack of knowledge of foreign markets Foreign market tariffs
Medical devices	Unfavorable tax treatment Availability of cores Healthcare costs	Transportation costs Regulatory barriers in foreign markets Lack of distribution or marketing channels
Retreaded tires	Availability of cores Energy costs Unfavorable tax treatment	([a])
Consumer products	Availability of low-cost new products Customer preferences for new products Declining demand for remanufactured goods	Lack of knowledge of foreign markets Transportation costs Lack of distribution or marketing channels
All other[b]	Healthcare costs Availability of cores Unfavorable tax treatment	Transportation costs Foreign market tariffs Lack of distribution or marketing channels
Wholesalers	Transportation costs Energy costs Labor costs	Regulatory barriers in foreign markets Transportation costs High price of cores

Source: USITC staff calculations of weighted responses to the Commission questionnaire.
[a]Neither foreign markets nor factors reported by survey respondents.
[b]Includes the electric apparatus, locomotives, office furniture, and restaurant equipment sectors.

U.S. producers also identify transportation costs as a major concern for both domestic and foreign markets. For example, some domestic HDOR equipment and motor vehicle parts remanufacturers have set up remanufacturing facilities in Mexico to take advantage of lower transportation costs owing to the country's close proximity to the U.S. market. Many of these same domestic remanufacturers also report lower labor costs in Mexico.

On the other hand, domestic tire remanufacturers (retreaders) have been most concerned with the scarcity of tire casings (cores), as surging demand for retreaded tires led to a casing shortage during 2009–11. Remanufacturers of aerospace products are chiefly concerned about a shortage of highly skilled technical labor.

The United States and Europe account for the bulk of global remanufacturing and trade, but other countries are developing their own remanufacturing industries.

Remanufacturing is well established in the United States and Europe. Although repair activities are common in a diverse range of industry sectors throughout the world, remanufacturing as a distinct industry is a relatively recent development in many countries, and trade in remanufactured goods by these countries is likely small. Large markets experiencing the greatest domestic demand for new aftermarket products also tend to be those that restrict trade in remanufactured goods the most. Such markets include Brazil, China, and India. The reasons given for such restrictions vary, but are often based on environmental, health, and consumer safety concerns, as well as perceptions of low quality. Other reasons reportedly include providing competitive advantages to, or protecting, competing domestic industries. In other countries or markets such as the EU, China, Singapore and the Republic of Korea, remanufacturing has been recognized by the government as supporting broader efforts to promote sustainability and reduce waste in industrial activities. With the exception of China, these countries are more open to trade in remanufactured goods.

Foreign direct investment (FDI) in U.S. remanufacturing supported over 22,000 full-time jobs in 2011; U.S. direct investment abroad was also substantial.

INTRODUCTION

U.S. remanufacturing employment associated with inbound FDI (investment in U.S. firms by foreign sources) accounted for 12 percent of total U.S. remanufacturing employment in 2011. Inbound FDI in the remanufactured machinery sector alone supported over 11,000 full-time jobs in the United States, while FDI in the U.S. aerospace and motor vehicle parts remanufacturing sectors supported a combined 8,500 full-time jobs. FDI in U.S. remanufacturing in 2011 totaled about $460 million, 80 percent of which went to the machinery remanufacturing sector. At the same time, U.S. direct investment flows to foreign remanufacturing totaled almost $300 million in 2011, mostly to the EU and Mexico. The aerospace, HDOR equipment, and IT products remanufacturing sectors accounted for about 80 percent of total U.S. outbound FDI in foreign remanufacturing.

Remanufacturing, or the industrial process of restoring end-of-life goods to original working condition, represents an important and growing segment of the U.S. manufacturing sector. In addition to offering “like new” functionality, remanufactured goods allow producers to considerably lessen their capital production costs and give consumers access to like-new products at lower prices than new goods.

Moreover, remanufacturing has lower environmental impacts than producing new goods, since it requires less material and energy. Remanufacturing occurs across a diverse range of U.S. industries and types of firms, including large original equipment manufacturers (OEMs), independent suppliers, and small and medium-sized enterprises (SMEs).

U.S. firms have been involved in remanufacturing for decades, and the United States is the leading global producer, consumer, and trader of remanufactured goods. However, despite the growing importance of remanufacturing, little information about U.S. remanufacturers or the U.S. and global markets for remanufactured goods is available.

Objective

As requested by the United States Trade Representative (USTR), [1] this report by the U.S. International Trade Commission (Commission or USITC) (1) provides an overview of the U.S. remanufactured goods industries and markets, (2) estimates U.S. and global trade in remanufactured goods, and (3) examines factors affecting trends in remanufactured goods trade. In addition, this report estimates and describes the size and scope of remanufacturing in the United States, including principal producers and levels of employment, investment, and sales; the U.S. market for remanufactured goods; U.S. exports of remanufactured goods, including exports to free trade agreement (FTA) partners; and the factors affecting sales, trade, and investment in U.S. remanufactured goods.

This report also includes an assessment of foreign direct investment (FDI) in U.S. remanufacturing, an overview of outward FDI (direct investment abroad) in remanufacturing by U.S.-based firms, an analysis of trade by foreign remanufacturers invested in the United States, and an overview of the global markets for remanufactured goods. This report covers the period 2009–11, as specified by the USTR.

Scope

The USTR asked that this report focus on remanufacturing-intensive sectors that account for the majority of remanufacturing activity in the United States. In his request, the USTR defined remanufactured goods as:

> non-agricultural goods that are entirely or partially comprised of parts that (i) have been obtained from the disassembly of used goods; and (ii) have been processed, cleaned, inspected, and tested to the extent necessary to ensure they have been restored to original working condition or better; and for which the remanufacturer has issued a warranty.

Approach

As requested by the USTR, most of the information presented in this report is based on the statistical analysis of primary data collected from responses to a Commission survey of U.S. firms believed to be remanufacturers (that survey is described in more detail below). This quantitative data was augmented by qualitative information developed as case studies on selected industry sectors and foreign markets for remanufactured goods.[2] The USTR specified that "this report should focus on remanufacturing-intensive sectors that account for the majority of remanufacturing activity in the United States." Based on interviews with industry representatives, the Commission identified 12 industry sectors that are believed to be remanufacturing-intensive (i.e., likely to include higher-than- average remanufacturing activity) and to account for the majority of remanufacturing activity in the United States (table 1.1). [3] In addition, Commission staff identified and included certain wholesaling activities in an effort to more accurately estimate trade in remanufactured goods by non-remanufacturing firms.

Information Sources

No databases that contain all known U.S. remanufacturing firms are known to exist. In addition, the Harmonized Tariff Schedule of the United States (HTS) does not generally distinguish between manufactured and remanufactured goods, and therefore, data on trade in remanufactured goods is largely unavailable.[4]

As a result, the Commission relied on multiple data sources to construct a database of firms that either remanufacture or are otherwise engaged in remanufacturing activities. Data sources included Commission staff research; a database of firms likely engaged in remanufacturing activities, obtained by Commission staff from the Rochester Institute of Technology (RIT); trade association membership lists of firms likely engaged in remanufacturing activities; ORBIS, a commercial database that compiles information on all known U.S. firms; and PIERS, a commercial database containing information on oceangoing import and export trade.

Commission Survey of U.S. Firms

The Commission sent questionnaires to a sample of U.S. firms identified in the database described above.[5] The questionnaire employed a stratified random sample to survey 7,000 U.S. firms, and Commission staff weighted results using two measures—firms' size as a proportion of the overall population, and the response rates of various categories of firms—to ensure that reported results accurately represented the entire population of remanufacturers.[6] Staff also used relative standard errors (RSEs) to gauge the precision of all data cited from the Commission's survey.

Box 1.1 provides a brief description of RSEs and how they are reported.

Table 1.1. Remanufacturing-intensive industry sectors in the United States

Sector	Product examples
Aerospace	Landing gear assemblies, flight surface actuators, engines, fuel systems
Consumer products	Consumer electronics, including mobile phones, cordless phones, smart phones, digital cameras, DVD players, televisions, etc.; and consumer household appliances, including washers, dryers, refrigerators, kitchen appliances
Electrical apparatus	Power distribution conductors, transformers, switch gears, and boards
Heavy-duty and off-road (HDOR) equipment	Diesel engines, transmissions, differential carriers, starters, brakes
Information technology (IT) products	Desktop computers, laptops, servers, data storage arrays, network switches, routers, modems, base transceiver stations, etc.; and imaging products, including copiers, printers, scanners, fax machines, ink and toner cartridges, fuser assemblies
Locomotives	Locomotive engines, drive motors, axles, other subsystems
Machinery	Industrial valves, turbines, machine tools, textile machinery, compressors
Medical devices	Diagnostic, surgical, and patient care apparatus; scanners; medical pumps; X-ray equipment
Motor vehicle parts	Starters, alternators, gear boxes, engines, differentials, brakes
Office furniture	Office systems furniture, laminate work surfaces, reupholstered office seating
Restaurant equipment	Ovens, beverage dispensers, food preparation tables
Retreaded tires	Tires for trucks, cars, off-the-road vehicles, airplanes

Source: Compiled by Commission staff.

Box 1.1. Measuring precision using relative standard errors (RSEs)

Throughout this report, all estimates based on calculations of weighted responses to the Commission's questionnaire have been examined to determine their precision. The RSE is a measure of the precision of these estimates that describes how widely the estimates are distributed around a mean. More specifically, an RSE is defined as the standard error of a particular estimate divided by the estimate itself. A smaller RSE indicates a more precise estimate and vice versa. For example, if an estimate of industry revenues is $10 million and the standard error for that estimate is $1 million, then the RSE is 0.10, or 10 percent. Likewise, if industry revenues are estimated at $10 million, but the standard error for that estimate is $5 million, then the RSE is 0.50, or 50 percent. The first example (with a smaller RSE) implies that the estimate is a more precise measurement of industry revenues, whereas the second example (with a larger RSE) suggests that another sample might produce a result different from the current estimate. Unless otherwise noted, estimates presented in this report have RSEs below 55 percent, which indicates that the standard error of the estimate is less than half of its magnitude. In cases where the survey produced an estimate that is particularly relevant to the reader but has less precision (i.e., a higher RSE), the RSE for that estimate was provided.[a] For instance, some estimates presented in this report have RSEs of 1.0, or 100 percent. In these cases, the standard errors for those estimates are greater than the value of the estimates themselves.

[a]For a more detailed description of RSEs, see appendix F.

The questionnaire asked firms to report remanufactured goods production, employment, investment, and trade for the period 2009–11. [7] The questionnaire also distinguished between firms that reported producing remanufactured goods and firms that reported performing remanufacturing activities other than production in order to capture remanufacturing employment and investment by both types of firms.[8] The questionnaire also identified foreign remanufacturers invested in the United States so that the Commission could perform the requested analysis of these firms' remanufacturingrelated trade, including trade in cores.[9] Limitations affecting the Commission's ability to estimate the size of remanufacturing activities are described in box 1.2.

Other Information Sources

This report draws on extensive qualitative information obtained from the public USITC hearing, written submissions, individual interviews, fieldwork, and U.S. embassy cable responses. The Commission held a public hearing on February 23, 2012. Witnesses included industry and trade association representatives covering a broad range of remanufacturing sectors, including the consumer products, heavy-duty and off-road (HDOR) equipment, information technology (IT) products, office furniture, medical devices, motor vehicle parts, and retreaded tires sectors, as well as an academic representative from RIT.[10] Written submissions were provided by a diverse group of trade associations and industry representatives. [11] The Commission conducted approximately 60 interviews with industry and academic representatives, as well as field work in Detroit, MI; Indianapolis, IN; Philadelphia, PA; Rochester, NY; and St. Louis, MO. Finally, to gather qualitative information on the global markets for remanufactured goods, the Commission sent information requests to U.S. embassies in 43 countries or economies,[12] and received responses from 33 of them.[13]

Box 1.2. Limitations on estimating the size of remanufacturing based on the Commission's questionnaire

Because of differences in terminology across industry sectors, ambiguity in the definition of remanufactured goods, and the nonsystematic survey of importers and firms with fewer than 20 employees, this report's estimates of remanufactured goods production and trade may be somewhat low.

Differences in terminology across industry sectors

Differences in terminology across industry sectors hinder efforts to accurately estimate remanufacturing activity because firms may not recognize themselves to be remanufacturers even though they perform remanufacturing activities as defined in this report. For instance, in the aerospace sector, maintenance and repair organizations (MROs) use the term "overhaul" to describe the remanufacturing of aircraft engines and airframes, but do not recognize the term "remanufacture." Likewise, in the motor vehicle parts sector, the term "rebuilding" may describe the remanufacturing of an item, although a firm might identify itself as a rebuilder, and not as a remanufacturer. Finally, most U.S. remanufacturers of medical imaging equipment identify themselves as refurbishers rather than remanufacturers because of the specific regulatory definition of "remanufacturer" provided by the U.S. Food and Drug Administration (FDA).[a] As a result, in each case some firms may have reported not producing remanufactured goods when their activities would likely fall within the scope of the definition of remanufactured goods provided by the USTR.

Ambiguity in the definition of remanufactured goods

Ambiguity in the definition of remanufactured goods also may have led to underreporting. For instance, some MROs that perform remanufacturing services on aircraft may not have equated providing these services with producing remanufactured goods. As a result, firms that provide remanufacturing services may not have reported producing remanufactured goods. Likewise, the extent of disassembly of cores into constituent parts varies by industry sector. Some firms, particularly in the IT sector, may not completely disassemble a product to be remanufactured, but would otherwise perform the activities to produce a remanufactured good as defined by the USTR. In this case, firms may or may not have reported producing remanufactured goods. Finally, firms that remanufacture goods for internal use may not provide a warranty on those particular products, and therefore report not producing remanufactured goods as defined by the USTR.

Nonsystematic survey of imports and firms with fewer than 20 employees

Estimates of U.S. production and imports of remanufactured goods during the 2009–11 period may be somewhat understated because the Commission did not systematically survey remanufacturing firms with fewer than 20 employees, or importers of remanufactured goods.

The Commission focused on those firms with 20 or more employees that were thought to account for the vast majority of U.S. remanufacturing employment and production of remanufactured goods.[b] Likewise, the Commission did not comprehensively survey all potential importers of remanufactured goods, as such a sample would need to include potential end users (including in nonremanufacturing-intensive sectors) and goods imported for personal consumption.

[a]For more information on the FDA's definition of remanufacturing, see chapter 7.

[b]Based on the statistical analysis of responses to the Commission questionnaire by the subset of the smallest SMEs (firms with 20–99 employees) and U.S. Census data, Commission staff estimated that there are approximately 5,000 additional remanufacturing firms with fewer than 20 employees. These firms were estimated to have accounted for 62 percent of U.S. remanufacturing firms, but for only an additional 4 percent ($1.7 billion) of U.S. production of remanufactured goods, and an additional 6 percent (11,600 full-time workers) of U.S. remanufacturing employment in 2011.

Overview of Remanufacturing Activities

The Remanufacturing Process

Remanufacturing is a manufacturing process that restores used or end-of-life products to original working condition or better ("like new") (figure 1.1). Remanufacturing occurs across a diverse range of industry sectors, but is more commonly concentrated among those sectors that produce capital-intensive, durable products with relatively longer product life cycles. As noted in table 1.1, remanufacturing-intensive sectors that are believed to account for the majority of remanufacturing activity in the United States include (in alphabetical order) aerospace, consumer products, electrical apparatus, HDOR equipment, IT products, locomotives, machinery, medical devices, motor vehicle parts, office furniture, restaurant equipment, and retreaded tires.

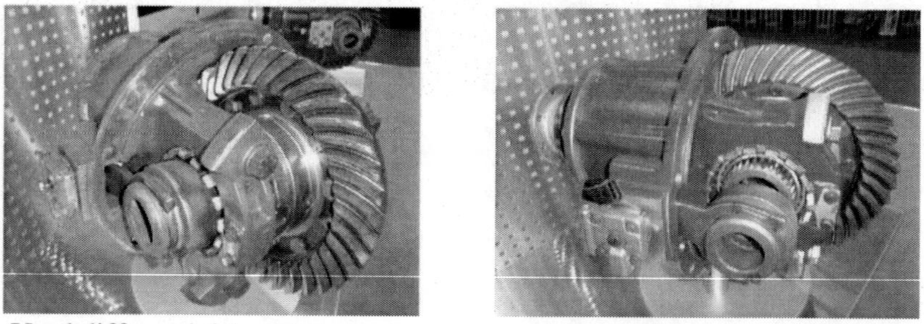

Used differential (core) Remanufactured differential

Source: USITC staff.

Note: A differential allows the wheels of a motor vehicle to rotate at different speeds when turning.

Figure 1.1. Remanufacturing restores used products to like-new condition.

The remanufacturing sectors described in this report use a variety of techniques to manage their supply of inputs, based largely on the concept of a closed-loop supply chain. Generally, unlike manufacturing processes that transform inbound virgin materials to make finished goods, the closed-loop supply chain in remanufacturing processes relies on used or discarded manufactured items (cores) as the principal material to be remanufactured (figure 1.2).

The remanufacturing process typically includes receiving inbound cores and assessing whether they are suitable to be remanufactured; disassembling, cleaning, and inspecting cores; replacing worn parts and restoring the core to original working condition or better; and then testing the remanufactured good.[14] Firms may be engaged in one or more (or all) of these activities. For example, a core broker may buy cores and disassemble them before selling the cores and constituent parts to a remanufacturer. Some firms may clean and inspect cores, but not actually remanufacture them. In addition, some firms may warehouse or distribute cores or remanufactured goods, some may trade (import and/or export) cores or remanufactured goods, while others may provide logistics and core management services to remanufacturers. As a result, the number and type of firms that comprise this remanufacturing "ecosystem" is substantially broader and more diverse than only those firms that produce remanufactured goods.

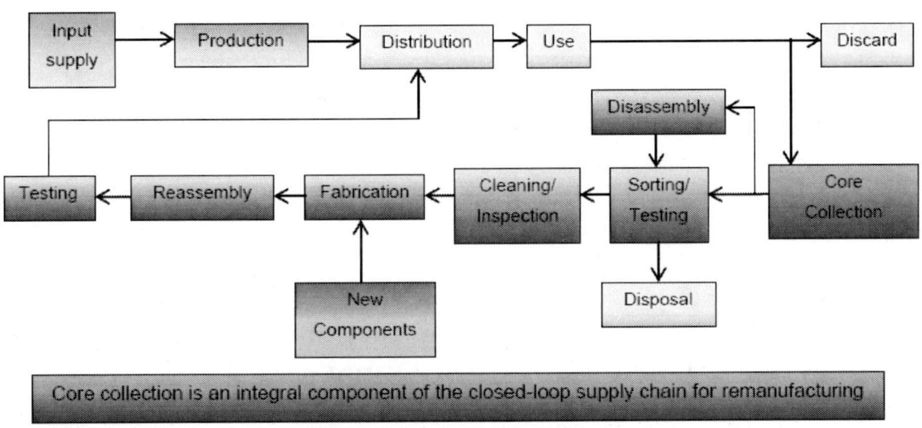

Source: Derived from Japke, Practice Guidelines, 2009, 10–11.

Figure 1.2. In contrast to manufacturing, remanufacturing relies on a closed-loop supply chain for cores.

Previous Estimates on the Size of U.S. Remanufacturing

According to industry representatives, the United States is the leading remanufacturer in the world; however, there are no definitive statistics to measure and compare the size of remanufacturing activities among countries. To date, efforts to estimate the size of remanufacturing activity or to estimate trade in remanufactured goods in the United States or elsewhere have been hampered by the absence of universally agreed upon and accepted definitions of "remanufacturing" and "remanufactured goods," as well as the absence of separate classifications for remanufactured goods in countries' economic and trade statistics. For example, the United States North American Industry Classification System (NAICS) does not distinguish between manufactured and remanufactured goods. As a consequence, basic economic statistics, including the number of establishments, the number of employees, and the value of industry shipments specifically for remanufactured goods in a given industry sector, are largely unavailable.[15]

The most exhaustive previous study on the size and scope of U.S. remanufacturing activity was conducted at Boston University in 1996.[16] Based on a sample of just over 1,000 firms in nine industry sectors comprising 83 major product areas,[17] the study estimated the total value of U.S. remanufacturing output to be $53 billion in 1996, with 73,000 firms engaged in remanufacturing activities, employing over 480,000 workers. Since then, however, the authors of the report have asserted that these estimates were overstated.[18]

A more recent study by the OEM Products-Services Institute (OPI) used a different method to estimate U.S. remanufacturing activity.[19] In that study, OPI estimated the share of remanufacturing expenditures for products in 16 broad industry sectors.[20] That study estimated total U.S. remanufacturing activity to be $49 billion in 2003, comparable to the size of remanufacturing activity estimated in the 1996 Boston University study. An updated assessment by OPI estimated that U.S. remanufacturing had risen to $92 billion in 2009.[21]

Organization of the Report

This report contains 10 chapters. Chapter 1 describes the report's objective, scope, and approach. Chapter 2 provides an overview of U.S. remanufacturing activities and provides information on remanufacturing industry sectors on an aggregated basis. Chapters 3–8 provide case studies for selected remanufacturing-intensive sectors that were chosen partly because they are large and more likely to export than other remanufacturing sectors, and partly because of their uniqueness compared with other remanufacturing sectors. These include (in order of production value from largest to smallest) the aerospace sector (chapter 3), the HDOR equipment sector (chapter 4), the motor vehicle parts sector (chapter 5), the IT products sector (chapter 6), the medical devices sector (chapter 7), and the retreaded tires sector (chapter 8). Chapter 9 surveys remanufacturing activities in six other sectors, including the consumer products, electrical apparatus, locomotives, machinery, office furniture, and restaurant equipment sectors.[22]

Finally, chapter 10 provides an overview of the global markets for remanufactured goods, including those in selected Asia-Pacific Economic Cooperation (APEC) and G-20 economies.[23] Information presented in chapters 2–9 responds to the first seven elements of the USTR's request. Information presented in chapter 10 responds to the eighth and final element of the USTR's request.

REFERENCES

Hauser, William, and Robert T. Lund. *The Remanufacturing Industry: Anatomy of a Giant.* Boston, MA: Boston University, 2003.

———. The Remanufacturing Industry: Hidden Giant, 1996. Boston, MA: Boston University, 1996.

———. Remanufacturing: Operating Practices and Strategies. Boston, MA: Boston University, 2008.

Giuntini, Ron, and Kevin Gaudette. "Remanufacturing: The Next Great Opportunity for Boosting U.S. Productivity." *Business Horizons,* November–December 2003, 41–48.

Japke, Onome. Practice Guidelines: Development of a Framework for Assessing the Economic Benefits of Remanufacturing. Cranfield, UK: Cranfield University, 2009.

OVERVIEW OF U.S. REMANUFACTURING ACTIVITIES

Introduction

This chapter provides an overview of the U.S. remanufacturing sector, including information on the size and scope of remanufacturing; the U.S. market for remanufactured goods; U.S. remanufacturing production, investment, and employment; U.S. trade in remanufactured goods and related cores; and foreign direct investment (FDI) in remanufacturing. This chapter aggregates and summarizes information presented in subsequent chapters of this report. The picture that emerges is of a broad segment of the U.S. economy characterized by diversity and growth in many sectors, driven by increasing demand for remanufactured goods in both domestic and foreign markets.

Size and Scope of Remanufacturing in the United States

Industry Sectors, Number of Firms, and Firm Size

As noted in chapter 1, remanufacturing-intensive sectors that are believed to account for the majority of remanufacturing activity in the United States include (in descending order of production value) aerospace, heavy duty and off-road (HDOR) equipment, motor vehicle parts, machinery, information technology (IT) products, medical devices, retreaded tires, and consumer products, as well as certain other sectors (electrical apparatus, locomotives, office furniture, and restaurant equipment).

Individual sectors with the greatest number of remanufacturing firms include machinery, IT products, motor vehicle parts, and aerospace (figure 2.1). Firms that sell, trade, warehouse, or distribute remanufactured goods or cores, but do not produce remanufactured goods, also account for a large number of remanufacturing firms and a notable portion of U.S. remanufacturing employment and trade in remanufactured goods. Remanufacturing sectors in which wholesalers are most active include motor vehicle parts, IT products, aerospace products, and to a lesser extent, consumer products. An estimated 2,900 firms with 20 or more employees remanufacture goods or perform some type of remanufacturing activity in the United States. There are several thousand additional firms with fewer than 20 employees that remanufacture goods or perform some type of remanufacturing activity. As noted in chapter 1 (box 1.1), the U.S. International Trade Commission (Commission) did not systematically survey remanufacturing firms with fewer than 20 employees, but estimated that there are approximately 5,000 of them in the United States. The Motor and Equipment Manufacturers Association (MEMA) estimates there are roughly 3,800 firms that remanufacture motor vehicle parts in the United States, whereas the Automotive Parts Remanufacturers Association (APRA) identifies more than 15,000 firms (most of which are small firms) that perform some of the remanufacturing steps on motor vehicle parts.[1] Small and medium-sized enterprises (SMEs)—defined as those firms that employ 20–499 workers—are estimated to account for an important share of remanufacturing employment, production, and trade. For instance, SMEs are estimated to have accounted for 36 percent (65,500 workers) of U.S. remanufacturing employment, 25 percent ($11.1 billion) of U.S. production of remanufactured goods, and 17 percent ($1.8 billion) of U.S. exports in 2011.[2]

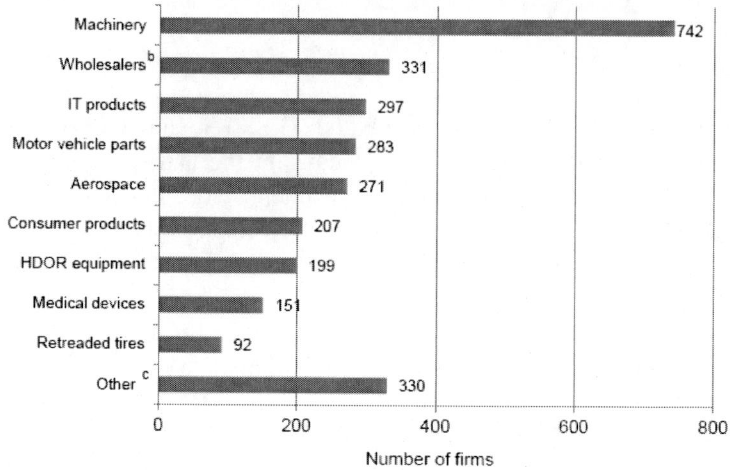

Source: USITC staff calculations of weighted responses to the Commission questionnaire.

[a] Firms with 20 or more employees.

[b] Firms that sell, trade (import or export), warehouse, or distribute remanufactured goods or cores, but do not produce remanufactured goods.

cIncludes the electrical apparatus, locomotives, office furniture, and restaurant equipment remanufacturing sectors.

Figure 2.1. Remanufactured goods: Estimated number of U.S. remanufacturers[a] by sector, 2011.

Production, Investment, and Employment

Production

Remanufactured goods make up a small but growing share of U.S. manufacturing output. Sales of these goods are estimated to have accounted for approximately 2 percent of total sales of all manufactured products by U.S. firms in these remanufacturing-intensive sectors during 2009–11 ($2.0 trillion for all manufactured products in these sectors in 2011).[3] Despite this relatively small size, remanufacturing is considered an integral and important part of company business plans. Further, U.S. production of remanufactured goods by these sectors is estimated to have increased by 15 percent from $37.3 billion in 2009 to $43.0 billion in 2011 (table 2.1).[4] The aerospace, HDOR equipment, and motor vehicle parts remanufacturing sectors are estimated to be the largest in terms of production value; however, the production value of remanufactured motor vehicle parts actually declined during the period (table 2.2). This decrease partly reflects falling demand for remanufactured motor vehicle parts as fleet customers delayed refreshing powertrains in their vehicles in an effort to cut costs, as noted in chapter 5.[5] By contrast, as described in chapter 3, the U.S. market for remanufactured aerospace products was bolstered by an aging aircraft fleet requiring continuous maintenance (and remanufacturing) services. U.S. production of remanufactured HDOR equipment, machinery, and retreaded tires grew the fastest during the study period.

Table 2.1. Remanufactured goods: U.S. production, investment, and employment, combined sectors, 2009–11

Type	2009	2010	2011	% change, 2009–11
		Thousand $		
Production	37,280,784	39,882,004	43,000,144	15.3
Investment				
In remanufacturing production	623,342	1,003,783	1,203,196	93.0
In non-production remanufacturing activities[a]	[b]15,805	19,563	20,129	27.3
Total investment	639,147	1,023,346	1,223,326	91.4
	Full-time workers			
Employment				
In remanufacturing production	154,682	158,656	165,776	7.2
In non-production remanufacturing activities	11,148	13,486	13,732	23.2
Total employment	165,830	172,142	179,508	8.2

Source: USITC staff calculations of weighted responses to the Commission questionnaire. Note: Totals may not sum due to rounding.

[a]Activities include disassembling cores into parts; cleaning, inspecting, and testing parts; and wholesale remanufacturing activities (selling, trading, warehousing, or distributing either cores or remanufactured goods).

[b]Low-precision estimate; RSE below 65 percent. RSEs are explained in chapter 1, box 1.1.

Table 2.2. Remanufactured goods: U.S. production of remanufactured goods by sector, 2009–11

Sector	2009	2010	2011	% change, 2009–11
		Thousand $		
Aerospace	11,691,316	12,677,916	13,045,513	11.6
HDOR equipment	5,152,938	6,244,302	7,770,586	50.8
Motor vehicle parts	7,018,303	6,969,823	6,211,838	–11.5
Machinery	4,059,570	4,774,291	5,795,105	42.8
IT products	2,709,170	2,592,831	2,681,603	–1.0
Medical devices	1,307,588	1,367,739	1,463,313	11.9
Retreaded tires	1,038,679	1,188,315	1,399,088	34.7
Consumer products	557,612	567,320	659,175	18.2
All other[a]	3,745,608	3,499,468	3,973,923	6.1
Total	37,280,784	39,882,004	43,000,144	15.3

Source: USITC staff calculations of weighted responses to the Commission questionnaire. Note: Totals may not sum due to rounding.

[a]The electrical apparatus, locomotive, office furniture, and restaurant equipment remanufacturing sectors.

Investment

U.S. remanufacturing investment is estimated to have nearly doubled during the study period, rising from $639 million in 2009 to $1.2 billion in 2011 (tables 2.1 and 2.3). The machinery, HDOR equipment, and motor vehicle parts sectors are estimated to account for the greatest share of, and growth in, remanufacturing investment. Remanufacturing investment in the aerospace and consumer products remanufacturing sectors, on the other hand, declined during the period.

Table 2.3. Remanufactured goods: U.S. remanufacturing investment by sector, 2009–11

Sector	2009	2010	2011	% change, 2009–11
		Thousand $		
Machinery	[a]206,578	[b]54,371	[b]711,008	244.2
HDOR equipment	[a]80,183	131,963	[c]162,746	103.0
Motor vehicle parts	76,106	105,669	105,684	38.9
Aerospace	95,777	77,379	90,471	–5.5
Medical devices	21,079	25,184	31,260	48.3
Retreaded tires	19,894	29,109	23,874	20.0
IT products	14,892	15,950	17,503	17.5
Consumer products	[a]34,035	7,416	4,948	–85.5
All other[d]	84,021	67,333	67,537	–19.6
Wholesalers	6,582	8,972	8,294	26.0
Total	639,147	1,023,346	1,223,326	91.4

Source: USITC staff calculations of weighted responses to the Commission questionnaire. Note: Totals may not sum due to rounding.

[a]Low-precision estimate; RSE below 80 percent.

[b]Low-precision estimate; RSE below 100 percent.

[c]Low-precision estimate; RSE below 65 percent.

[d]The electrical apparatus, locomotive, office furniture, and restaurant equipment remanufacturing sectors.

Employment

U.S. remanufacturing employment is estimated to have increased by 8 percent from about 166,000 workers in 2009 to about 180,000 workers in 2011 (tables 2.1 and 2.4). Workers that produce remanufactured goods are estimated to account for over 90 percent of U.S. remanufacturing employment. The aerospace, motor vehicle parts, and machinery sectors are estimated to account for the greatest share of remanufacturing employment, although HDOR equipment and IT products remanufacturers account for a sizable portion of employment as well. Remanufacturing employment in the IT products, wholesalers, and retreaded tires sectors experienced the largest relative growth during the period. Indeed, an estimated 60 percent of tire retreaders upgraded and expanded retreading operations in response to continued growth in the U.S. economy, as noted in chapter 8.

U.S. Market for Remanufactured Goods

Market Size

The U.S. market for remanufactured goods is the largest in the world, and it is estimated to have increased by 15 percent from \$36.0 billion in 2009 to \$41.5 billion in 2011 (table 2.5). The United States is a net exporter of remanufactured goods, although imports are estimated to have accounted for about one-quarter of the U.S. market in 2011, up from one-sixth in 2009.[6] However, domestic production grew during the same period, and U.S. remanufacturers exported an increasingly larger share of domestic production, driven primarily by growth in wholesaler and remanufactured HDOR equipment exports.

Table 2.4. Remanufactured goods: U.S. remanufacturing employment, by sector, 2009–11

Sector	2009	2010	2011	% change, 2009–11
		Full-time workers		
Aerospace	35,804	36,349	35,201	–1.7
Motor vehicle parts	30,069	30,404	30,653	1.9
Machinery	24,814	24,870	26,843	8.2
HDOR equipment	18,072	18,654	20,870	15.5
IT products	11,493	13,025	15,442	34.4
Wholesalers	8,180	10,619	10,891	33.1
Consumer products	8,216	7,733	7,613	–7.3
Retreaded tires	3,935	4,368	4,880	24.0
Medical devices	3,781	3,910	4,117	8.9
All other[a]	21,467	22,210	22,999	7.1
Total	165,830	172,142	179,508	8.2

Source: USITC staff calculations of weighted responses to the Commission questionnaire. Note: Totals may not sum due to rounding.

[a]The electrical apparatus, locomotive, office furniture, and restaurant equipment remanufacturing sectors.

Table 2.5. Remanufactured goods: U.S. market (apparent consumption) for remanufactured goods, 2009–11

Type	2009	2010	2011	% change, 2009–11
		Thousand $		
Production	37,280,784	39,882,004	43,000,144	15.3
Imports	6,256,042	9,508,060	10,263,056	64.1
Exports	7,502,991	8,805,067	11,735,665	56.4
U.S. apparent consumption	36,033,835	40,584,997	41,527,535	15.2

Sources: USITC staff calculations of weighted responses to the Commission questionnaire. Figures for U.S. imports and exports of retreaded tires are taken from official statistics of the U.S. Department of Commerce (USDOC) for Harmonized Tariff Schedule of the United States (HTS) subheadings 4012.11, 4012.12, 4012.13, and 4012.19.

Note: Totals may not sum due to rounding.

U.S. Buyers and Demand Factors

U.S. Buyers

In general, buyers of remanufactured goods overlap substantially with buyers of new products. Purchases of remanufactured goods in many of the industry sectors described in subsequent chapters of this report tend to be “business to business” transactions. In other words, buyers tend to be firms that, in turn, provide goods and services to end users. For instance, as noted in chapter 3, buyers of remanufactured aerospace components include aircraft manufacturers, airline operators and subsidiaries, and independent third-party suppliers that provide remanufacturing parts and services for aircraft. Commercial fleet owners (e.g., logistics or trucking firms) are major drivers of demand for remanufactured motor vehicle parts, since they have large numbers of vehicles and since the low-cost maintenance of their vehicles is a priority, as described in chapter 5. Original equipment manufacturers (OEMs) also purchase remanufactured motor vehicle parts to reduce the cost of filling warranty claims and to fulfill their obligations to provide replacement parts for older vehicles. Retreaded tires are purchased primarily by truck, bus, aviation, construction, and mining equipment operators. Hospitals, surgical centers, imaging centers, and private care facilities constitute the principal consumers of remanufactured medical devices. Finally, Amtrak and freight railroad operators (particularly the larger, or Class I railroads) comprise the leading customers for locomotive systems.

The U.S. armed services are also important U.S. buyers of remanufactured goods, having purchased an estimated 13 percent ($5.5 billion) of the remanufactured goods produced in the United States in 2011.[7] An estimated 65 percent ($3.6 billion) of U.S. armed services purchases were remanufactured aerospace products, while 23 percent ($1.3 billion) were remanufactured motor vehicle parts. In fact, the U.S. armed services are estimated to have purchased 27 percent of all aerospace products remanufactured in the United States, and 21 percent of all remanufactured motor vehicle parts.[8]

Demand Factors

The price of remanufactured goods versus new goods, including the availability of lower-priced new alternatives, and consumer perceptions about price and quality primarily influence demand for remanufactured goods in the U.S. market.

All things being equal, remanufactured goods are often viewed as economical alternatives to new products because they are functionally equivalent and, on average, are priced less than a comparable new good.

As described in later chapters of this report, remanufactured goods typically save consumers 30–50 percent (or more) on the price of an equivalent new product. For instance, the average reported price of a retreaded truck tire is about 40 percent less than the price of a new one, as noted in chapter 8. By using retreads instead of new tires, the trucking industry reportedly saves more than $3 billion annually. Remanufactured motor vehicle parts reportedly cost on average 20–50 percent less than most new aftermarket parts. Remanufactured medical devices can cost end users 50– 75 percent less than a comparable new device; cost-cutting measures and the general shift toward value-driven healthcare increased demand for these goods.

On the other hand, the availability of low-cost alternatives, particularly imports of new products that are priced comparably to remanufactured goods, can dampen the demand for remanufactured goods. For instance, U.S. imports of new low-cost motor vehicle parts, particularly from Asia, may substitute for domestic remanufactured vehicle components, thereby reducing domestic producers' sales of remanufactured aftermarket products and potentially cutting domestic jobs, as described in chapter 5. Likewise, the cost of new (and often imported) IT products, such as personal computers (PCs) or printers, have been declining relative to remanufactured products.

Since IT products often have shorter life cycles, consumers may prefer to switch to newer, better, and less costly products instead of buying remanufactured IT goods, as noted in chapter 6. For these types of goods, customers need to believe that the remanufactured product will offer excellent functionality as well as an attractive price. The likelihood of consumers choosing remanufactured goods instead of new ones varies, depending on the product in question as well as buyers' preferences.

U.S. Suppliers and Supply Factors

U.S. Suppliers

Sellers of remanufactured goods include OEMs that remanufacture and sell their own new products, independent remanufacturers, and wholesalers and retailers. However, the specific constellation of sellers varies by sector. In the aerospace sector, OEMs and maintenance and repair organizations (MROs), including aircraft operators, their subsidiaries, and independent service providers, supply remanufactured goods and services for aircraft overhaul and maintenance. The major sellers of remanufactured HDOR equipment tend to be the OEMs themselves, along with their affiliates and licensed distributors. Similarly, the sale of remanufactured locomotives is dominated by the half-dozen major locomotive manufacturers. Suppliers of remanufactured motor vehicle parts are more diverse, and include OEMs (through their dealerships and distribution centers), independent remanufacturers, retail stores, and independent repair shops. Finally, unaffiliated distributors, wholesalers, or warehousing entities sell remanufactured goods in a wide range of sectors, accounting for a sizable portion of U.S. sales and trade.

Supply Factors

A variety of factors influence the supply of remanufactured goods in the U.S. market, including (1) the availability and cost of cores; (2) other manufacturing costs, including transportation and labor; (3) the availability of skilled labor; (4) the relationship between OEMs and independent remanufacturers; and (5) investment in research and development (R&D). Factors cited by U.S. remanufacturers as affecting their ability to compete in the U.S. market are shown in figure 2.2.[9]

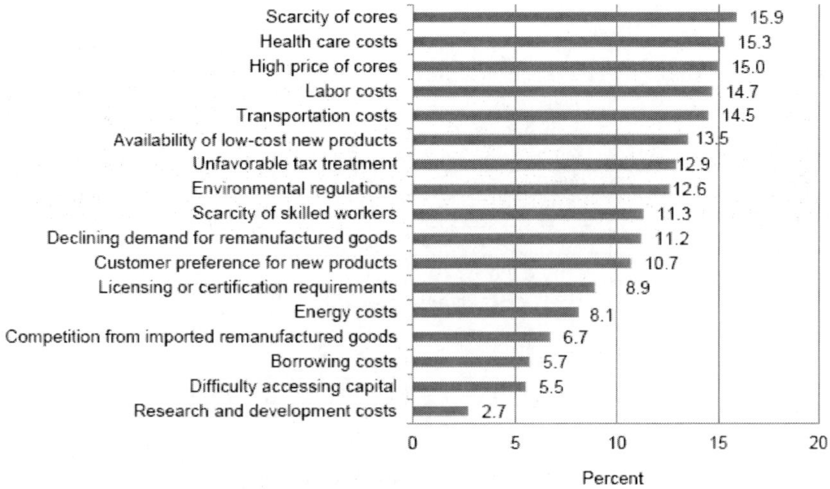

Source: USITC staff calculations of weighted responses to the Commission questionnaire.

Note: More than one factor could be cited by U.S. remanufacturers as "extremely important." As a result, shares do not total to 100 percent.

Figure 2.2. Remanufactured goods: Competitive factors cited as "extremely important," by percent of U.S. remanufacturers.

Factors considered to be the most important vary across sectors, reflecting the diversity among sectors and how factors affect them differently (table 2.6). For instance, over 50 percent of U.S. tire retreaders cited scarcity of cores as an "extremely important" factor affecting their competitive position, whereas only 6 percent of machinery remanufacturers reported this as an "extremely important" factor. Indeed, as noted in chapter 8, increasing U.S. demand for retreaded tires in certain tire segments in 2010 and 2011 reportedly led to a shortage of used tire casings.

Supply of Cores

The supply of cores to be remanufactured is limited at the beginning of the life cycle of a new product (such as the introduction of a new model), generally increases as the product type or model matures, and finally declines as the product type or model begins to be withdrawn from the market. During this cycle, warranty coverage tends to increase the flow of cores as buyers opt to replace rather than repair worn-out goods. For instance, as noted in chapter 5, when a component of a new engine is first introduced in a motor vehicle, few to no cores are available to remanufacture that part. As the part ages, however, consumers replace the component and return an increasing number of cores, thereby creating a supply of cores viable for remanufacturing. Core availability begins to decline again as the vehicles in which the part is used are removed from service. Remanufacturers use a variety of core management techniques to ensure an adequate supply of cores, based largely on the concept of a closed-loop supply chain (box 2.1).

Table 2.6. Remanufactured goods: Top competitive factors cited as "extremely important," by share of U.S. remanufacturers, by sector

Sector	Factor	Percent
Retreaded tires	Availability of cores	52.3
HDOR equipment	Transportation costs	36.4
Motor vehicle parts	Labor costs	33.0
Consumer products	Availability of low-cost new products	31.1
IT products	High price of cores	24.4
Wholesalers	Transportation costs	23.1
Machinery	Availability of skilled workers	22.1
Aerospace	Availability of skilled workers	19.5
Medical devices	Unfavorable tax treatment	[a]15.5

Source: USITC staff calculations of weighted responses to the Commission questionnaire.
[a]Low-precision estimate; RSE below 65 percent.

Box 2.1. Core management and the closed-loop supply chain

The remanufacturing sectors described in subsequent chapters of this report use a variety of core management techniques, based largely on the concept of a closed-loop supply chain. Remanufactured goods are often sold with an additional core deposit price incorporated into the price of the good.

Once a core is returned to the remanufacturer (not necessarily the original manufacturer of the product), the deposit is credited back to the customer who returned the core. Such a system incentivizes core returns and ensures an adequate and reliable source of cores.

For instance, some of the largest OEM remanufacturers of motor vehicle parts and

HDOR equipment use a closed-loop system in which an OEM's own distribution system collects cores and returns them to the OEM's service divisions. Similarly, larger OEM medical device remanufacturers work closely with their own distribution networks to procure cores, and some independent remanufacturers will even become "preferred vendors" of OEMs in order to secure their supply of medical device cores. In contrast, some independent remanufacturers may rely on purchases on the open market to fill needed core supplies. For instance, independent motor vehicle parts remanufacturers may work with third parties such as brokers, salvage operators, and salvage auctions to procure an adequate supply of cores.

Transportation and Labor Costs

Transportation and labor costs strongly influence where remanufactured goods are produced. Over 36 percent of HDOR equipment remanufacturers and 32 percent of motor vehicle parts remanufacturers cited transportation costs as an "extremely" important competitive factor, reflecting in part the importance of regional production networks and producing closer to the intended market of sale.[10] The costs of shipping inbound cores to be processed and outbound remanufactured goods for distribution contribute to the tendency of remanufacturing to be done close to the U.S. market, favoring Canada and Mexico. This is particularly true for motor vehicle parts and IT products, as noted in chapters 5 and 6. Some domestic IT products remanufacturers have invested in remanufacturing operations in Mexico to serve the U.S. market, and have taken advantage of lower labor costs and lower transportation costs owing to the country's close proximity to the United States. Leading U.S. motor vehicle parts remanufacturers, such as Cardone Industries and Remy, also have remanufacturing facilities in Mexico that supply the U.S. market.

Availability of Skilled Workers

A scarcity of skilled workers can limit the supply of remanufactured goods, and was cited as the top concern for the aerospace and machinery remanufacturing sectors. As noted in chapter 3, the supply of services and parts from aerospace MROs is closely correlated with the availability of skilled and certified maintenance and repair technicians. Because the skills needed to provide maintenance services and to remanufacture parts are highly technical and require much experience and training, the pool of qualified workers is limited. MROs are constantly seeking to hire qualified workers and/or to train new ones when possible.

Relationship between OEM and Independent Remanufacturers

The relationship between OEMs and independent remanufacturers can influence the supply of remanufactured goods, particularly in the IT field. For instance, as noted in chapter 6, some independent printer cartridge remanufacturers claim that OEMs engage in various tactics to compete with third-party firms, including public relations campaigns to discredit the quality of third-party remanufactured printer cartridges and efforts to limit the supply of cores. These efforts reportedly include making cartridges more technologically complex and thus more difficult for third-party firms to remanufacture, and promoting discount programs directly with customers to exchange new cartridges at a discount for used ones. In other remanufacturing sectors, OEMs and independent remanufacturers may work more collaboratively and on a contractual basis, as in the remanufactured medical device sector (described in chapter 7).

Investment in R&D

Often remanufacturers must conduct extensive R&D to maintain remanufacturing capabilities and improve their remanufacturing processes. For instance, as noted in chapter 5, many independent motor vehicle parts remanufacturers do not have the original product design schematics; their ability to re-engineer and find solutions to problems with original components influences the supply and quality of remanufactured motor vehicle parts. One area that requires extensive R&D investment is the remanufacturing of electronic components or components with embedded elec-tronics, and remanufacturers have increasingly hired engineers with specialized knowledge in both mechanical and electrical engineering to improve remanufacturing capabilities in these areas.

Estimates of U.S. Trade in Remanufactured Goods and Cores

Trade in Remanufactured Goods

Exports of Remanufactured Goods

Between 2009 and 2011, U.S. exports of remanufactured goods are estimated to have increased by more than 50 percent, from \$7.5 billion in 2009 to \$11.7 billion in 2011 (table 2.7). Nevertheless, U.S. exports of remanufactured goods are estimated to have accounted for only 3 percent of total U.S. exports of goods (new and remanufactured) in these remanufacturing-intensive sectors during the 2009–11 period (\$407.3 billion in 2011).[11] Wholesalers (i.e., firms that do not produce remanufactured goods, but sell or trade them) account for nearly one-third of U.S. exports of remanufactured goods.[12] Exports of remanufactured aerospace products, HDOR equipment, and machinery are estimated to account for the largest shares of U.S. exports of remanufactured goods, and exports of remanufactured HDOR equipment and machinery are also among the fastest-growing export sectors, due in large part to growing global demand for equipment used in construction, mining, and energy exploration. Some remanufacturing sectors export far more of their production than others. Although remanufactured motor vehicle parts ranks as the third-largest remanufacturing sector in the United States in terms of production value (see table 2.2), the sector accounted for only about 5 percent of total U.S. exports of remanufactured goods in 2011. The remanufactured motor vehicle sector primarily produces and sells remanufactured parts for the domestic market, and trade in remanufactured goods and cores tends to reflect cross-border production networks linking the United States and neighboring Canada or Mexico, or both. In contrast, HDOR equipment remanufacturers are estimated to export 30 percent of their production, and are thus more directly affected by foreign barriers to trade in remanufactured goods. Canada, the European Union (EU), and Mexico are estimated to be the largest markets for U.S. exports of remanufactured goods (table 2.8). Between 2009 and 2011, U.S. exports of these goods to North American Free Trade Agreement (NAFTA) partners Canada and Mexico are estimated to have increased by 45 percent from \$2.1 billion in 2009 to \$3.1 billion in 2011, accounting for 27 percent of the U.S. total for that year, and for 71 percent of U.S. exports of remanufactured goods to free trade agreement (FTA) partners (table 2.9). U.S. exports of these goods to FTA partners are estimated to have increased by 49 percent from \$3.0 billion in 2009 to \$4.4 billion in 2011, accounting for almost 40 percent of the U.S. total for that year. In comparison, U.S. exports to non-FTA countries are estimated to have increased by about 60 percent from $4.6 billion in 2009 to $7.3 billion in 2011.

Table 2.7. Remanufactured goods: U.S. exports of remanufactured goods by sector, 2009–11

Sector	2009	2010	2011	% change, 2009–11
		Thousand $		
Wholesalers	1,139,996	1,965,778	3,751,538	229.1
Aerospace	2,546,579	2,525,582	2,589,543	1.7
HDOR equipment	[a]1,695,950	[a]2,045,076	[a]2,451,967	44.6
Machinery	858,281	884,716	1,348,734	57.1
Motor vehicle parts	430,145	494,145	581,520	35.2
Medical devices	453,770	470,858	488,008	7.5
IT products	219,961	250,197	260,032	18.2
Consumer products	[b]13,733	[a]16,608	[a]21,151	54.0
Retreaded tires	15,904	16,495	18,545	16.6
All other[c]	128,672	135,613	224,627	74.6
Total	7,502,991	8,805,067	11,735,665	56.4

Sources: USITC staff calculations of weighted responses to the Commission questionnaire. Figures for U.S. exports of retreaded tires are taken from official statistics of the USDOC for HTS subheadings 4012.11, 4012.12, 4012.13, and 4012.19.

Note: Totals may not sum due to rounding.

[a]Low-precision estimate; RSE below 65 percent.

[b]Low-precision estimate; RSE below 80 percent.

[c]The electrical apparatus, locomotive, office furniture, and restaurant equipment remanufacturing sectors.

Table 2.8. Remanufactured goods: U.S. exports of remanufactured goods by leading destination, combined sectors, 2009–11

Country or market	2009	2010	2011	% change, 2009–11
		Thousand $		
Canada	1,419,597	1,808,022	2,263,811	59.5
EU	1,013,104	1,063,457	1,413,458	39.5
Mexico	724,554	471,158	854,201	17.9
Australia	269,049	356,264	441,229	64.0
Singapore	198,231	305,134	417,490	110.6
Indonesia	[a]151,465	160,039	[a]192,361	27.0
China	127,245	161,235	175,965	38.3
Brazil	70,367	107,057	136,606	94.1
Chile	110,362	137,978	127,858	15.9
Japan	169,637	107,148	117,680	–30.6
All other countries or markets	3,249,380	4,127,576	5,595,006	72.2
Total	7,502,991	8,805,067	11,735,665	56.4

Sources: USITC staff calculations of weighted responses to the Commission questionnaire. Figures for U.S. exports of retreaded tires are taken from official statistics of the USDOC for HTS subheadings 4012.11, 4012.12, 4012.13, and 4012.19.

Note: Totals may not sum due to rounding.

[a]Low-precision estimate; RSE below 65 percent.

Table 2.9. Remanufactured goods: U.S. exports of remanufactured goods to FTA partners, combined sectors, 2009–11

FTA partners	2009	2010	2011	% change, 2009–11
		Thousand $		
Canada	1,419,597	1,808,022	2,263,811	59.5
Mexico	724,554	471,158	854,201	17.9
Australia	269,179	356,499	441,378	64.0
Singapore	198,247	305,140	417,557	110.6
Chile	110,481	138,259	127,886	15.8
CAFTA-DR[a]	[b]71,597	47,984	[c]110,105	53.8
Peru	19,265	36,650	[c]100,967	424.1
Israel	80,300	63,358	51,267	–36.2
Oman	24,247	38,843	31,444	29.7
Morocco	17,185	9,876	10,576	–38.5
Jordan	8,470	11,245	7,941	–6.2
Bahrain	3,253	3,942	5,668	74.2
Total	2,946,375	3,290,978	4,422,799	50.1

Sources: USITC staff calculations of weighted responses to the Commission questionnaire. Figures for U.S. exports of retreaded tires are taken from official statistics of the USDOC for HTS subheadings 4012.11, 4012.12, 4012.13, and 4012.19.

Note: Totals may not sum due to rounding.

[a]Dominican Republic-Central America-United States Free Trade Agreement partners.

[b]Low-precision estimate; RSE below 65 percent.

[c]Low-precision estimate; RSE below 80 percent.

After Canada and Mexico, Australia was estimated to be the third-largest FTA partner market for U.S. remanufactured goods. As noted in chapter 4, Australia is a large market for U.S. exports of remanufactured HDOR equipment used in the mining industry. U.S. exports of remanufactured HDOR equipment are estimated to account for over two-thirds of total U.S. exports of remanufactured goods to Australia.

U.S. exports of remanufactured goods by foreign remanufacturers invested in the United States[13] are estimated to have increased by 63 percent from $1.2 billion in 2009 to $2.0 billion in 2011, accounting for 17 percent of total U.S. exports that year (table 2.10). The machinery and wholesale sectors are estimated to have accounted for about 80 percent of such exports in 2011.

The aerospace sector is estimated to have accounted for 13 percent of such exports in 2011, and likely reflects EU aviation companies invested in the United States, as noted in chapter 3. Both NAFTA partners and the EU are important markets for these types of U.S. exports of remanufactured goods (table 2.11).

Imports of Remanufactured Goods

U.S. imports of remanufactured goods are estimated to have increased by 64 percent from $6.3 billion in 2009 to $10.3 billion in 2011 (table 2.12). The IT products, aerospace, HDOR equipment, and motor vehicle parts sectors are estimated to have accounted for the vast majority of U.S. imports of remanufactured goods during the period. Mexico is the largest supplier of remanufactured goods to the U.S. market, accounting for about 30 percent ($3.1 billion) of total U.S. remanufactured imports in 2011 (table 2.13).

Table 2.10. Remanufactured goods: U.S. exports of remanufactured goods by foreign remanufacturers invested in the United States by sector, 2009–11

Sector	2009	2010	2011	% change, 2009–11
		Thousand $		
Machinery	[a]732,464	667,682	[a]1,120,608	53.0
Wholesalers	117,670	[a]460,217	[a]450,944	283.2
Aerospace	250,788	310,602	253,402	1.0
Motor vehicle parts	24,632	[b]53,305	[b]71,745	191.3
Medical devices	79,175	71,015	65,569	–17.2
HDOR equipment	[a]8,204	[a]15,902	[a]26,364	221.4
IT products	16,575	19,067	18,692	12.8
All other[c]	[a]2,294	[a]2,106	[a]2,013	–12.2
Total	1,231,801	1,599,896	2,009,337	63.1

Source: USITC staff calculations of weighted responses to the Commission questionnaire.

Note: Figures for U.S. exports of retreaded tires by foreign remanufacturers invested in the United States are not available. Totals may not sum due to rounding.

[a]Low-precision estimate; RSE below 65 percent.

[b]Low-precision estimate; RSE below 80 percent.

[c]The electrical apparatus, locomotive, office furniture, and restaurant equipment remanufacturing sectors.

Table 2.11. Remanufactured goods: U.S. exports of remanufactured goods by foreign manufacturers invested in United States by leading destination, 2009–11

Country or market	2009	2010	2011	% change, 2009–11
		Thousand $		
Mexico	[a]388,182	[b]147,272	[a]426,102	9.8
EU	293,276	284,835	288,073	–1.8
Canada	116,302	161,699	187,893	61.6
All other countries or markets	434,041	1,006,090	1,107,270	155.1
Total exports	1,231,801	1,599,896	2,009,337	63.1

Source: USITC staff calculations of weighted responses to the Commission questionnaire.

Note: Figures for U.S. exports of retreaded tires by foreign remanufacturers invested in the United States are not available. Figures for "all other countries" are calculated. Totals may not sum due to rounding.

[a]Low-precision estimate; RSE below 100 percent.

[b]Low-precision estimate; RSE below 80 percent.

As noted in chapter 6, in the IT products sector, Mexico plays a particularly important role in computer hardware remanufacturing, since its proximity to the United States gives it a significant transportation cost advantage. While Mexico is also an important assembly location for the manufacturing of certain new IT products, its role is even more prominent in remanufacturing, owing mainly to the complexities of core procurement and the greater importance of transportation cost factors in remanufacturing. In the aerospace sector, as noted in chapter 3, the EU is an important trading partner for remanufactured goods, reflecting the integration of aircraft maintenance and remanufacturing services between the U.S. and EU markets. In contrast, imports of retreaded tires are negligible because U.S. retreaded tire production has historically supplied almost all of domestic consumption, as noted in chapter 8.

Table 2.12. Remanufactured goods: U.S. imports of remanufactured goods by sector, 2009–11

Sector	2009	2010	2011 % change, 2009–11	
		Thousand $		
IT products	1,931,994	2,281,198	2,756,475	42.7
Aerospace	484,156	[a]1,992,524	1,869,901	286.2
Wholesalers	960,364	1,435,328	1,874,128	95.1
HDOR equipment	1,042,284	1,302,481	1,489,259	42.9
Motor vehicle parts	1,219,922	1,462,269	1,481,939	21.5
Consumer products	[b]325,095	[a]394,162	[b]360,264	10.8
Machinery	[c]136,460	[c]504,861	[c]268,256	96.6
Medical devices	109,776	85,036	110,705	0.8
Retreaded tires	6,014	8,513	11,446	90.3
All other[d]	39,977	41,688	40,683	1.8
Total	6,256,042	9,508,060	10,263,056	64.1

Sources: USITC staff calculations of weighted responses to the Commission questionnaire. Figures for U.S. imports of retreaded tires are taken from official statistics of the USDOC for HTS subheadings 4012.11, 4012.12, 4012.13, and 4012.19.

Note: Totals may not sum due to rounding.

[a]Low-precision estimate; RSE below 65 percent.

[b]Low-precision estimate; RSE below 80 percent.

[c]Low-precision estimate; RSE below 100 percent.

[d]The electrical apparatus, locomotive, office furniture, and restaurant equipment remanufacturing sectors.

Table 2.13. Remanufactured goods: U.S. imports of remanufactured goods by leading source, combined sectors, 2011

Country or market	2011	% share
	Thousand $	
Mexico	3,058,680	29.8
EU	819,503	8.0
China	[a]412,078	4.0
All other countries or markets	5,972,795	58.2
Total	10,263,056	100.0

Sources: USITC staff calculations of weighted responses to the Commission questionnaire. U.S. imports of retreaded tires from official statistics of the USDOC for HTS subheadings 4012.11, 4012.12, 4012.13, and 4012.19.

Note: Totals may not sum due to rounding.

[a]Low-precision estimate; RSE below 65 percent.

U.S. imports of remanufactured goods by foreign remanufacturers invested in the United States are estimated to have increased by 50 percent from $1.0 billion in 2009 to $1.5 billion in 2011 (table 2.14), accounting for 15 percent of total U.S. imports of remanufactured goods that year. Remanufactured motor vehicle parts made up the largest category by far; they are estimated to have accounted for about 43 percent ($664 million) of total U.S. imports of remanufactured goods by foreign remanufacturers invested in the United States in 2011. The EU was a leading source of U.S. imports of remanufactured goods by foreign remanufacturers invested in the United States (table 2.15), but sector trade patterns were quite diverse.

Remanufactured aerospace products are estimated to have accounted for 70 percent ($212 million) of these types of U.S. imports from the EU, likely reflecting EU aviation companies invested in (and importing products into) the United States. Wholesalers are estimated to have accounted for around 90 percent ($73 million) of these U.S. imports from China.

Table 2.14. Remanufactured goods: U.S. imports of remanufactured goods by foreign remanufacturers invested in the United States by sector, 2009–11

Sector	2009	2010	2011	% change, 2009–11
		Thousand $		
Motor vehicle parts	[a]469,880	[a]598,759	[a]664,402	41.4
Wholesalers	[b]120,173	[b]185,367	[b]283,328	135.8
Aerospace	282,308	204,562	281,643	–0.2
Machinery	[a]135,066	[a]500,795	[a]263,182	94.9
HDOR equipment	([d])	([d])	([d])	([d])
IT products	([d])	([d])	([d])	([d])
Medical devices	[c]20,459	[c]22,964	[c]31,865	55.8
Total	1,029,549	1,522,597	1,539,395	49.5

Source: USITC staff calculations of weighted responses to the Commission questionnaire.

Note: Figures for U.S. imports of retreaded tires, consumer products, and "all other" remanufacturing sectors (i.e., electrical apparatus, locomotive, office furniture, and restaurant equipment) by foreign remanufacturers invested in the United States are not available. Totals may not sum due to rounding.

[a]Low-precision estimate; RSE below 100 percent.

[b]Low-precision estimate; RSE below 80 percent.

[c]Low-precision estimate; RSE below 65 percent.

[d]Suppressed to prevent the disclosure of confidential business information (CBI).

Table 2.15. Remanufactured goods: U.S. imports of remanufactured goods by foreign remanufacturers invested in the United States by leading source, combined sectors, 2011

	2011	% share
Country or market	Thousand $	
EU	306,621	20.0
China	[a]77,716	5.0
All other countries or markets	1,155,058	75.0
Total	1,539,395	100.0

Source: USITC staff calculations of weighted responses to the Commission questionnaire.

Note: Figures for U.S. imports of retreaded tires by foreign remanufacturers invested in the United States are not available. There were no reported U.S. imports of remanufactured goods by foreign remanufacturers invested in the United States in the consumer products and "all other" (electrical apparatus, locomotive, office furniture, and restaurant equipment) remanufacturing sectors. Totals may not sum due to rounding.

[a]Low-precision estimate; RSE below 65 percent.

Trade in Cores

Core Exports

The United States is a net exporter of cores. U.S. core exports are estimated to have increased by 38 percent from $1.5 billion in 2009 to $2.0 billion in 2011 (table 2.16). The aerospace, IT products, and HDOR remanufacturing sectors are estimated to account for the majority of U.S. core exports. In contrast, U.S. exports of tire casings were negligible, as the vast majority of tire casings are recycled back into downstream U.S. retreading operations intended for domestic consumption, as noted in chapter 8.

Table 2.16. Remanufactured goods: U.S. core exports by sector, 2009–11

Sector	2009	2010	2011	% change, 2009–11
		Thousand $		
Aerospace	383,175	488,385	599,801	56.5
IT products	504,381	579,341	507,153	0.5
HDOR equipment	239,814	254,346	256,230	6.8
Consumer products	[a]14,405	[a]67,429	[a]198,322	1,276.8
Motor vehicle parts	[b]147,847	[a]177,587	[a]195,144	32.0
Machinery	[b]52,646	[a]81,312	[a]115,949	120.2
Medical devices	96,622	70,639	87,860	–9.1
Wholesalers	42,181	53,662	80,727	91.4
Retreaded tires	29,259	32,993	39,288	34.3
All other	[a]45	[a]1,640	[a]2,738	5,984.4
Total	1,510,372	1,807,333	2,083,211	37.9

Sources: USITC staff calculations of weighted responses to the Commission questionnaire. Figures for U.S. exports of retreaded tire cores (used pneumatic tires, including tire casings) are taken from official statistics of the USDOC for HTS subheading 4012.20.

Note: Used pneumatic tires is a basket category that includes tire casings suitable for retreading, as well as other used tires that may not be suitable for retreading. Totals may not sum due to rounding.

[a]Low-precision estimate; RSE below 80 percent.

[b]Low-precision estimate; RSE below 65 percent.

Mexico is the largest market for U.S. core exports, and is estimated to have accounted for 43 percent ($885 million) of total U.S. core exports in 2011 (table 2.17). For instance, as noted in chapter 6, some U.S.-based IT products remanufacturers follow a regional production model and export IT cores to Mexico to be remanufactured there, and then import finished remanufactured goods.

Table 2.17. Remanufactured goods: U.S. core exports by leading destination, combined sectors, 2011

Country or market	2011 Thousand $	% share
Mexico	899,200	43.2
China	[a]224,967	10.8
EU	208,773	10.0
Other countries or markets	750,270	36.0
Total	2,083,211	100.0

Sources: USITC staff calculations of weighted responses to the Commission questionnaire. Figures for U.S. exports of retreaded tire cores (used pneumatic tires, including tire casings) are taken from official statistics of the USDOC for HTS subheading 4012.20.

Note: Totals may not sum due to rounding.

[a]Low-precision estimate; RSE below 100 percent.

Table 2.18. Remanufactured goods: U.S. core exports by foreign remanufacturers invested in the United States by sector, 2009–11

Sector	2009	2010	2011	% change, 2009–11
		Thousand $		
Motor vehicle parts	[a]85,440	[a]110,510	[a]122,760	43.7
Machinery	[b]47,570	[b]70,931	[b]100,207	110.7
Aerospace	[b]22,043	[c]36,080	[c]39,070	77.2
Wholesalers	[b]18,444	[b]24,622	[b]30,851	67.3
HDOR equipment	([d])	([d])	([d])	([d])
IT products	([d])	([d])	([d])	([d])
Medical devices	13,786	11,092	16,462	19.4
Total	187,691	254,780	310,776	65.6

Source: USITC staff calculations of weighted responses to the Commission questionnaire.

Note: Figures for U.S. exports of retreaded tire cores (tire casings) by foreign remanufacturers invested in the United States are not available. There were no reported U.S. core exports by foreign remanufacturers invested in the United States for the consumer products and "all other" (electrical apparatus, locomotive, office furniture, and restaurant equipment) remanufacturing sectors. Totals may not sum due to rounding.

[a]Low-precision estimate; RSE below 100 percent.

[b]Low-precision estimate; RSE below 80 percent.

[c]Low-precision estimate; RSE below 65 percent.

[d]Suppressed to prevent the disclosure of CBI.

Table 2.19. Remanufactured goods: U.S. core exports by foreign remanufacturers invested in the United States by leading destination, combined sectors, 2011

	2011	
Country or market	Thousand $	% share
EU	58,838	18.9
Mexico	[a]30,666	9.9
Singapore	[b]28,049	9.0
All other countries or markets	193,223	62.2
Total	310,776	100.0

Source: USITC staff calculations of weighted responses to the Commission questionnaire. Note: Totals may not sum due to rounding.

[a]Low-precision estimate; RSE below 80 percent.

[b]Low-precision estimate; RSE below 100 percent.

The prevalence of this model is supported by the importance of Mexico as an import source for finished remanufactured goods, as noted previously. The same is likely true for remanufactured HDOR equipment, as an estimated 98 percent of HDOR core exports are destined for FTA partners, as noted in chapter 4.

U.S. core exports by foreign remanufacturers invested in the United States are estimated to have accounted for 15 percent ($311 million) of total U.S. core exports in 2011. Cores of motor vehicle parts and machinery are estimated to have accounted for about 72 percent of such U.S. core exports in 2011 (table 2.18). The EU, Mexico, and Singapore were important markets for U.S. core exports by foreign remanufacturers invested in the United States (table 2.19).

Core Imports

U.S. imports of cores are estimated to have more than doubled from $797 million in 2009 to $1.8 billion in 2011 (table 2.20). Wholesalers are estimated to have imported the greatest share of cores, followed by the HDOR equipment and machinery remanufacturing sectors. The EU is estimated to have supplied almost half of total U.S. core imports in 2011 (table 2.21). Canada and Australia are much smaller core suppliers to the U.S. market; nevertheless, one U.S. remanufacturer of motor vehicle parts and HDOR equipment noted that it shipped a substantial amount of cores from Australia to the United States.[14]

Table 2.20. Remanufactured goods: U.S. core imports by sector, 2009–11

Sector	2009	2010	2011	% change, 2009–11
		Thousand $		
Wholesalers	[a]103,806	[a]258,176	[a]638,530	515.1
HDOR equipment	[b]488,651	[b]521,318	618,622	26.6
Machinery	10,335	[b]14,263	[c]122,205	1,082.4
Medical devices	[b]67,419	[b]84,754	[a]116,489	72.8
Aerospace	43,203	50,122	94,348	118.4
Motor vehicle parts	20,222	50,903	57,756	185.6
Retreaded tires	24,430	29,171	43,858	79.5
IT products	19,275	23,986	33,823	75.5
All other[d]	[b]19,640	23,268	10,026	–49.0
Total	796,981	1,055,960	1,735,656	117.8

Sources: USITC staff calculations of weighted responses to the Commission questionnaire. Figures for U.S. imports of retreaded tire cores (used pneumatic tires, including tire casings) are taken from official statistics of the USDOC for HTS subheading 4012.20.

Note: Used pneumatic tires is a basket category that includes tire casings suitable for retreading, as well as other used tires that may not be suitable for retreading. Totals may not sum due to rounding.

[a]Low-precision estimate; RSE below 80 percent.

[b]Low-precision estimate; RSE below 65 percent.

[c]Low-precision estimate; RSE below 100 percent.

[d]The electrical apparatus, locomotive, office furniture, and restaurant equipment remanufacturing sectors.

Table 2.21. Remanufactured goods: U.S. core imports by leading source, combined sectors, 2011

Country or market	2011	% share
	Thousand $	
EU	825,776	47.6
Canada	184,016	10.6
Australia	[a]113,913	6.6
All other countries or markets	611,951	35.3
Total	1,735,656	100.0

Sources: USITC staff calculations of weighted responses to the Commission questionnaire. Figures for U.S. imports of retreaded tire cores (used pneumatic tires, including tire casings) are taken from official statistics of the USDOC for HTS subheading 4012.20.

Note: Totals may not sum due to rounding.

[a]Low-precision estimate; RSE below 65 percent.

Table 2.22. Remanufactured goods: U.S. core imports by foreign remanufacturers invested in the United States by sector, 2009–11

Sector	2009	2010	2011	% change, 2009–11
		Thousand $		
Machinery	9,386	[a]12,538	[b]118,934	1,167.1
HDOR equipment	([c])	([c])	([c])	([c])
IT products	[d]1,186	[d]1,582	[d]3,164	166.8
Medical devices	([c])	([c])	([c])	([c])
Total	22,910	27,901	142,130	520.4

Source: USITC staff calculations of weighted responses to the Commission questionnaire.

Note: Figures for U.S. imports of retreaded tire cores (tire casings) by foreign remanufacturers invested in the United States are not available. There were no reported U.S. core exports by foreign remanufacturers invested in the United States for the aerospace, motor vehicle parts, consumer products, “all other” (electrical apparatus, locomotive, office furniture, and restaurant equipment), and wholesalers sectors. Totals may not sum due to rounding.

[a]Low-precision estimate; RSE below 65 percent.

[b]Low-precision estimate; RSE below 100 percent.

[c]Suppressed to prevent the disclosure of CBI.

[d]Low-precision estimate; RSE below 80 percent.

U.S. core imports by foreign remanufacturers invested in the United States are estimated to have accounted for 8 percent ($142 million) of total U.S. core imports in 2011, and are primarily machinery cores (table 2.22).

European remanufacturers invested in the United States are important core suppliers to the U.S. market.

Factors Affecting Trade in Remanufactured Goods

Most of the factors affecting the competitive position of domestic remanufacturers in the U.S. market also affect those competing in foreign markets.

Major concerns common to both markets include transportation costs and the high price of cores; transportation costs ranked as the leading factor affecting the ability of U.S. remanufacturers to compete in foreign markets (figure 2.3).[15]

Import prohibitions/bans, foreign regulatory barriers, and tariffs were also cited as “extremely important” factors.

In addition, the absence of a commonly accepted legal definition of remanufactured goods was cited by witnesses at the Commission hearing as a principal barrier to trade in these products.

According to hearing witnesses, both the product and the remanufacturing process are poorly understood by many customs authorities, who often view remanufactured goods as used goods, junk goods, or simply as waste.[16]

As a consequence, some countries, including Brazil, China, and India, restrict or prohibit trade in remanufactured goods or cores. In contrast, several U.S. FTAs contain provisions that define and address remanufactured goods for the purposes of establishing rules of origin and clarifying the nondiscriminating treatment of and market access for remanufactured goods.[17]

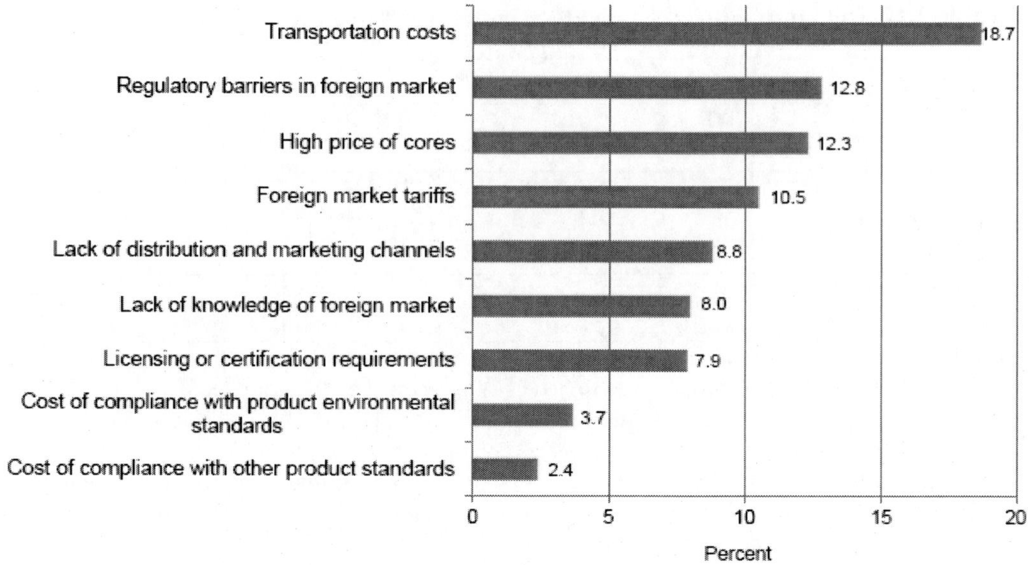

Source: USITC staff calculations of weighted responses to the Commission questionnaire.

Figure 2.3. Remanufactured goods: Competitive factors cited by U.S. remanufacturers as "extremely important" that affect their ability to compete in foreign markets, by share of U.S. remanufacturers that export.

Foreign Direct Investment in Remanufacturing[18]

Inbound FBI in U.S. Remanufacturing Activities

Between 2009 and 2011, FDI flows in U.S. remanufacturing activities are estimated to have increased by 65 percent, from \$277 million in 2009 to \$458 million in 2011 (table 2.23). As noted in chapter 9, the remanufactured machinery sector is estimated to have received over 80 percent (\$378 million) of foreign remanufacturing investment in 2011, likely reflecting a foreign acquisition or other large investment in U.S. remanufacturing operations by one or two firms during the period. Likewise, the large estimated decline in foreign investment in the remanufactured motor vehicle parts sector likely reflects a foreign acquisition of U.S. remanufacturing operations by one or two firms in 2009. Important foreign sources of FDI in U.S. remanufacturing activities include Japan and the EU. Firms from Switzerland, China, Mexico, Canada, India, and Singapore have also invested in U.S. remanufacturing activities.

Foreign firms invested in U.S. remanufacturing activities are an important source of U.S. remanufacturing employment.

Between 2009 and 2011, annual U.S. employment attributable to these firms is estimated to be around 22,000 workers, or 12 percent of total U.S. remanufacturing employment (table 2.24). The remanufactured machinery sector is estimated to account for almost one-half of U.S. remanufacturing employment associated with FDI in U.S. remanufacturing activities.

As described in chapter 3, the aerospace sector is estimated to account for 22 percent of U.S. remanufacturing employment associated with FDI, likely reflecting EU aviation firms invested in MROs located in the United States.

Table 2.23. Remanufactured goods: FDI in U.S. remanufacturing by sector, 2009–11

Sector	2009	2010	2011	% change, 2009–11
		Thousand $		
Machinery	[a]22,955	[b]414,512	[b]377,901	1,546.3
Aerospace	[c]33,699	[c]27,193	[c]41,051	21.8
Motor vehicle parts	[c]201,904	40,084	[c]15,237	-92.5
Medical devices	[a]11,361	[a]13,085	[a]11,962	5.3
IT products	([d])	[c]1,572	([d])	n/a
All other[e]	[c]7,024	[c]10,690	[c]11,380	62.0
Total	276,944	[a]507,136	[b]457,532	65.2

Source: USITC staff calculations of weighted responses to the Commission questionnaire. Note: Totals may not sum due to rounding. N/a = not applicable.

[a]Low-precision estimate; RSE below 80 percent.

[b]Low-precision estimate; RSE below 100 percent.

[c]Low-precision estimate; RSE below 65 percent.

[d]None reported.

[e]Includes the consumer products, electrical apparatus, locomotive, office furniture, and restaurant equipment remanufacturing sectors.

Table 2.24. Remanufactured goods: U.S. remanufacturing employment associated with inbound FDI, 2009–11

Sector	2009	2010	2011	% change, 2009–11
		Full-time workers		
Machinery	10,102	10,262	11,077	9.7
Aerospace	4,403	4,648	4,872	10.7
Motor vehicle parts	3,541	3,748	3,576	1.0
IT products	[a]2,457	[a]2,199	[b]1,799	-26.8
Medical devices	410	395	375	-8.5
HDOR equipment	259	135	173	-33.2
Wholesalers	29	32	60	106.9
Retreaded tires	([c])	([c])	([c])	n/a
All other[d]	263	267	271	3.0
Total	21,465	21,686	22,203	3.4

Source: USITC staff calculations of weighted responses to the Commission questionnaire. Note: Totals may not sum due to rounding. N/a = not applicable.

[a]Low-precision estimate; RSE below 80 percent.

[b]Low-precision estimate; RSE below 65 percent.

[c]None reported.

[d]The electrical apparatus, locomotive, office furniture, and restaurant equipment remanufacturing sectors.

Outbound FBI in Foreign Remanufacturing Activities

Between 2009 and 2011, U.S. direct investment flows in foreign remanufacturing activities are estimated to have increased by 141 percent from $119 million in 2009 to $287 million in 2011 (table 2.25). Investment in foreign HDOR remanufacturing activities is estimated to have accounted for almost 40 percent ($111 million) of total U.S. investment in foreign remanufacturing activities in 2011.

As noted in chapter 4, some of the largest U.S.-based HDOR equipment remanufacturers, including Caterpillar, Cummins, and John Deere, have remanufacturing operations located in or near markets experiencing growth in demand for aftermarket parts, including in Brazil, China, and India. In May 2011, Caterpillar opened a remanufacturing facility in Singapore as the regional source for remanufactured components to serve Asia's mining industries.

Other reasons for investing in foreign markets include barriers to imports of remanufactured goods and cores.

Table 2.25. Remanufactured goods: U.S. outbound FDI in foreign remanufacturing activities by sector, 2009–11

Sector	2009	2010	2011	% change, 2009–11
		Thousand $		
HDOR equipment	36,505	99,719	111,267	204.8
IT products	[a]410	[a]202	[a]70,966	17,208.8
Aerospace	11,468	70,758	47,708	316.0
Wholesalers	[b]24,972	[b]25,110	[b]26,199	4.9
Motor vehicle parts	11,567	9,244	13,616	17.7
Medical devices	[a]6,577	[a]7,608	[b]10,992	67.1
Machinery	5,080	2,213	4,397	–13.4
Consumer products	[a]1,766	[b]955	599	–66.1
All other[c]	20,872	10,103	1,421	–93.2
Total	119,219	225,914	287,166	140.9

Source: USITC staff calculations of weighted responses to the Commission questionnaire. Note: Totals may not sum due to rounding.

[a]Low-precision estimate; RSE below 80 percent.

[b]Low-precision estimate; RSE below 65 percent.

[c]The electrical apparatus, locomotive, office furniture, and restaurant equipment remanufacturing sectors.

Table 2.26. Remanufactured goods: U.S. outbound FDI in foreign remanufacturing activities by leading destination, combined sectors, 2009–11

Country or market	2009	2010	2011	% change, 2009–11
		Thousand $		
EU	9,958	60,802	105,069	955.1
Mexico	50,552	75,758	71,817	42.1
All other countries or markets	58,710	89,354	110,280	87.8
Total	119,219	225,914	287,166	140.9

Source: USITC staff calculations of weighted responses to the Commission questionnaire.
Note: Figures for "all other countries or markets" are calculated. Totals may not sum due to rounding.

The EU is estimated to have received 37 percent ($105 million) of U.S. FDI in foreign remanufacturing activities (table 2.26). Mexico is estimated to have accounted for 25 percent ($72 million) of U.S. FDI in foreign remanufacturing activities. As previously noted, U.S. remanufacturers have invested in Mexican remanufacturing facilities to take advantage of lower labor costs and lower transportation costs owing to their proximity to the U.S. market.

AEROSPACE

Description of Aerospace Remanufacturing Activities

The aerospace remanufacturing sector consists of firms that remanufacture aircraft components or provide remanufacturing services on aircraft and their subsystems. [1] Airplanes and their subsystems account for the bulk of remanufactured products in the aerospace industry. [2] The main systems and subsystems of an airplane that are remanufactured include the airframe, [3] engines, avionics, hydraulics, and interior furnishings.[4] Other aircraft components that are typically remanufactured include wheels and brakes, auxiliary power units, fuel systems, flight controls, thrust reversers, landing gear assemblies, and electrical systems.[5] Globally, the aerospace remanufacturing sector is driven in large part by the mandatory aircraft safety inspections prescribed by national aircraft certification authorities (e.g., the Federal Aviation Administration, or FAA, in the United States). Firms that produce remanufactured aerospace parts or components or provide remanufacturing services on aircraft are part of a larger aircraft maintenance industry, generally referred to as maintenance and repair organizations (MROs). MROs repair existing aircraft parts to a certain standard, or replace existing parts with either new or remanufactured parts (or serviceable parts) to restore the aircraft to "airworthy" condition (box 3.1).

Box 3.1. Maintenance and repair organizations

Maintenance and repair organizations (MROs) are firms that overhaul engines, perform heavy maintenance on airframes, repair components, and carry out line maintenance on aircraft to restore aircraft to airworthy condition. (Line maintenance is performed at the gate or on the ramp between flights.) Airlines are their main customers, and every aircraft must be inspected on a regular basis to assure its airworthiness. Normally, remanufacturing activities in the aerospace sector involve engine overhaul, airframe maintenance, and component repair. Line maintenance typically does not involve any remanufacturing. In addition to remanufacturing components and parts of an aircraft, MROs may also install new parts to restore the aircraft to airworthy condition. Remanufacturing activities are thus only a part of the larger market for aircraft maintenance and repair. MROs are licensed by a country's aircraft certification agency, such as the FAA in the United States and the European Aviation Safety Agency (EASA) in Europe. They adhere to standards set forth by both the governmental agency and the aircraft's OEM for aircraft and component life, and perform the mandated inspections required to assure the airworthiness of an aircraft. An aircraft is airworthy if all its parts, components, systems, and structures conform to its type design and it is in a condition for safe operation.[a] MROs will accept an aircraft and perform the necessary disassembly of the airframe and systems based on the owner/operator's contract with them. Typically, MROs either remanufacture any needed replacement parts or subcontract with specialty shops to restore the part to airworthiness. MROs then reinstall the remanufactured part onto the aircraft. MROs therefore provide a removal and replacement service in addition to remanufacturing the part (or subcontracting its remanufacturing).

[a] U.S. Department of Transportation, FAA, 14 C.F.R., part 3, sec. 3.5.

Parts that need to be replaced (either because they do not meet safety standards or because their use is time limited) can be replaced either by new parts or parts that have been repaired to a condition that meets the safety standard set for that part. For the purposes of this report, aerospace parts that are repaired to a condition that meets the safety standard set for that part as determined by a given governmental certification agency, such as the FAA,[6] are considered to meet the definition of remanufactured goods that was provided in the USTR's request letter.[7]

The remanufacturing of aircraft parts primarily involves overhaul activities to restore a component to airworthiness. In fact, the term "remanufacture" rarely appears in the aerospace industry—instead, the terms "overhaul," "rebuild," and "component repair" are generally used to denote remanufacturing. In addition, "rotables" and "serviceable parts" commonly refer to remanufactured aerospace parts that replace the original part or component.[8] Nevertheless, for the purposes of this report, the terms "remanufactured goods" and "remanufacturing" will be used, where appropriate, in lieu of these terms.

Size of Aerospace Remanufacturing in the United States

Industry Structure

In the United States, close to 4,200 firms reportedly perform maintenance and repair activities, although not all of them remanufacture aerospace products or provide remanufacturing services. SMEs reportedly make up 85 percent of all such firms and account for 21 percent of all U.S. employment in MRO activities.[9]

The majority of these companies are concentrated in California (780 firms), Florida (523), and Texas (392).[10] A variety of firms—OEMs, airline operators and their maintenance subsidiaries, MROs, and independent suppliers—perform remanufacturing activities in the aerospace industry. Any firm that has been approved by a country's aviation certification agency to repair a part is considered a "repair station."

Repair stations can be large OEMs or airline operators that have been approved to remanufacture or make a range of repairs to many different parts, or they can be small firms that have been granted approval to remanufacture or make repairs to a single part. Often, large MROs will have networks of small specialty repair suppliers to whom they might contract out specific parts to be remanufactured.

As noted in box 3.1, remanufacturing activities in the aerospace sector are concentrated in engine overhaul, airframe heavy maintenance, and component repair. Generally, OEMs, airline operators and their maintenance subsidiaries, and independent suppliers or MROs perform engine overhaul and heavy maintenance, while smaller specialty repair service providers typically specialize in the remanufacture of specific aircraft components. Specialty repair service providers are authorized by aviation safety authorities to conduct specific types of repairs to parts and components, and often provide contract remanufacturing services on aerospace parts and components to MROs (figure 3.1).[11]

It is estimated that almost all remanufacturers in the aerospace sector remanufacture products that exceed the original working condition of the parts they overhaul.[12]

In addition, an estimated 68 percent of remanufacturers also produce new products, while 88 percent also sell new goods to customers.[13]

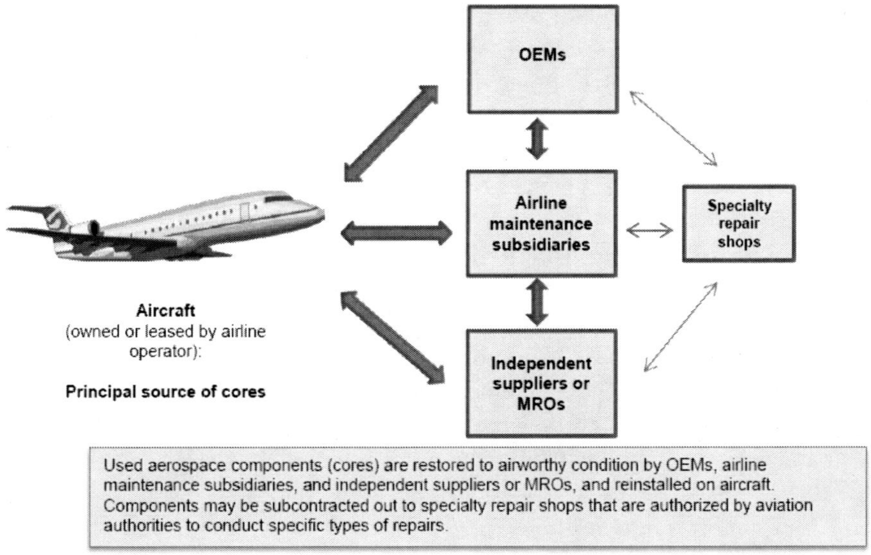

Source: Commission staff.

Figure 3.1. The supply chain for remanufactured aircraft parts involves a number of industry participants.

Engine Overhaul

Engine overhaul reportedly accounts for the bulk of aerospace remanufacturing activities.[14] OEMs are the largest providers of engine overhaul services, followed by airline operators and their subsidiaries, and independent suppliers. Globally, the most significant and well-known OEMs that provide engine overhaul services are General Electric, Pratt & Whitney, and Rolls-Royce.

Major airline operators that overhaul their own engines include United Airlines, Japan Airlines, and Iberia Airlines. [15] Airline subsidiaries that provide engine overhaul services to other airline operators include Air France Industries/ KLM Engineering and Maintenance (French and Dutch ownership), Delta TechOps (U.S. ownership), and Lufthansa Technik Group (German ownership).[16]

Independent suppliers that provide engine overhaul services include Standard Aero, MTU Aero Engines (German and U.S. ownership), Timco, Pacific Gas Turbine Center, LLC (U.S. subsidiary of SR Technics Switzerland AG and Chromalloy Gas Turbine LLC), and ST Aerospace (Swiss ownership).

Airframe Heavy Maintenance

Under safety and operational guidelines established by the aircraft manufacturer and the national aviation authorities, an aircraft's airframe must undergo inspection and repair at certain time intervals. Airline operators and their subsidiaries typically perform much of the heavy airframe maintenance, with the rest performed by independent, third-party contractors. Some airlines, such as British Airways, American Airlines, and Japan Airlines, perform a majority of maintenance themselves. Airline subsidiaries, such as Air France Industries/KLM Engineering & Maintenance, Delta TechOps, and Lufthansa Technik Group, perform airframe heavy maintenance for both their parent airlines and other airline operators. Some of the larger independent suppliers that provide heavy airframe maintenance services to airline operators include AAR Corp., Aviation Technical Services (ATS), SR Technics, Timco (U.S. ownership), Hong Kong Aircraft Engineering Ltd. (HAECO) (Hong Kong ownership), and ST Aerospace.[17] Although OEM involvement in airframe heavy maintenance has traditionally been relatively limited, many of the largest airframe OEMs, such as Boeing and Airbus, have created services organizations to capture more of this market.

Component Repair and Remanufacturing

Compared with the engine overhaul and airframe heavy maintenance segments, the component repair and remanufacturing segment is more fragmented, with more SMEs involved. Again, however, OEMs, airline operators and their maintenance subsidiaries, and independent suppliers all either remanufacture aircraft components or provide remanufacturing services, although the share of activities performed by each type of firm is reportedly more evenly split among them than in the other two industry segments.[18]

Production, Investment, and Employment

Between 2009 and 2011, the production of remanufactured aerospace products is estimated to have risen by 12 percent from $11.7 billion in 2009 to $13.0 billion in 2011 (table 3.1). Despite this increase, remanufactured aerospace products are estimated to have accounted for less than 3 percent of total sales by U.S. firms within the aerospace sector during that time period ($504 billion in 2011).[19] At the same time, U.S. investment in aerospace remanufacturing is estimated to have fallen slightly from $96 million in 2009 to $90 million in 2011. Employment in aerospace remanufacturing is estimated to have remained around 35,800 full-time workers over the same period.

Table 3.1. Aerospace: Remanufacturing production, investment, and employment, 2009–11

Item	2009	2010	2011	% change, 2009–11
		Thousand $		
Production	11,691,316	12,677,916	13,045,513	11.6
Investment	95,777	77,379	90,471	–5.5
		Full-time workers		
Employment	35,804	36,349	35,201	–1.7

Source: USITC staff calculations of weighted responses to the Commission questionnaire.

U.S. Market for Remanufactured Aerospace Products

Market Size

Between 2009 and 2011, the U.S. market for remanufactured aerospace products is estimated to have increased by 28 percent from $9.6 billion in 2009 to $12.3 billion in 2011 (table 3.2). In comparison, although estimates vary, the broader North American market for maintenance and repair services is estimated at between $16.6 billion and $39 billion, which is roughly one-third to one-half of the global market for maintenance and repair activities.[20]

The United States is a net exporter of remanufactured aerospace products, making the size of the U.S. market slightly smaller than domestic production. Exports are important for the U.S. industry, accounting for about 20 percent of domestic production during the period, while imports accounted for 15 percent of the U.S. market in 2011, up from 5 percent in 2009.

Table 3.2. Aerospace: U.S. market (apparent consumption) for remanufactured goods, 2009–11

Item	2009	2010	2011	% change, 2009–11
		Thousand $		
Production	11,691,316	12,677,916	13,045,513	11.6
Imports	484,156	1,992,524	1,869,901	286.2
Exports	2,546,579	2,525,582	2,589,543	1.7
U.S. apparent consumption	9,628,893	12,144,857	12,325,872	28.0

Source: USITC staff calculations of weighted responses to the Commission questionnaire. Note: Totals may not sum due to rounding.

Demand for remanufactured aerospace products and maintenance and repair services is influenced primarily by the strict regulatory environment for aircraft safety. Other factors influencing demand are competition among airlines, leading them to cut costs; the age of the existing aircraft fleet; and the growth of the new fleet. Factors affecting the supply of remanufactured aerospace components and remanufacturing services include advances in technology and design, OEMs expanding their service offerings to include remanufacturing services, and the availability of skilled labor.

U.S. Buyers and Demand Factors

Buyers of remanufactured aerospace components include OEMs, airline operators and maintenance subsidiaries, MROs, and independent third-party suppliers that provide remanufactured parts and remanufacturing services on aircraft. In order to meet safety certification requirements, every part of an aircraft must be regularly inspected, and parts that have outlived their time-in-place date or fail to meet OEM specifications must be replaced. Competition among airlines has led them to press MROs to cut prices on these mandatory maintenance and repair services. Measures to reduce costs have created demand for a vast pool of rotables that meet the authorities' safety standards as a less costly alternative to new parts. The age of the aircraft fleet and the type of service (daily versus occasional, long haul versus short haul) influence the demand for remanufactured aerospace products and the extent of servicing and maintenance. Generally, the older the fleet, the heavier its servicing and maintenance requirements. The United States has one of the oldest airline fleets in the world, with major carriers' aircraft averaging over 12 years old. In contrast, major European fleets average 9–13 years old, while major Asian air carriers' fleets average 7–11 years old.[21] According to industry estimates, the North American maintenance and repair market and, by extension, the market for remanufactured aerospace products is estimated to grow more slowly than in other regions in the world.[22] One reason for this is that North America, like Europe, is considered a mature market. Most of the growth in demand over the next 10 years is forecast to occur in emerging markets such as the Asia-Pacific region and the Middle

East.[23] However, according to market projections, North America will still possess one of the largest fleets of large civil aircraft and remain a significant market for remanufactured aerospace parts. [24] Both Boeing and Airbus project robust demand for air travel and aircraft over the next 20 years.[25] As a result, there will likely be a significant increase in the number of aircraft that will require regular inspection.

U.S. Sellers and Supply Factors

OEMs, airline maintenance subsidiaries, MROs, and independent suppliers sell remanufactured aerospace products or provide remanufacturing services. Surplus dealers may also sell remanufactured aerospace components to MROs. Advances in technology and design may limit the supply of remanufactured aerospace products; new aircraft are increasingly designed to require less maintenance. The estimated costs and complexity of servicing the components and structures on the newer aircraft will be a challenge to the established repair industry, as new techniques and machines will be necessary to service more advanced aircraft.[26] The entry of OEMs into the maintenance and repair market has introduced greater competition into the market. OEMs traditionally have had only limited involvement in airframe heavy maintenance. However, as the maintenance and repair market has grown, the largest airframe OEMs, Boeing and Airbus, have created services to capture more of this market. For example, in February 2012 Boeing announced the launch of Boeing Edge, characterized as a one-stop service shop for airlines that operate Boeing planes.[27] Boeing has also launched its GoldCare program, in which Boeing will lead a team of maintenance and repair partners to maintain and maximize performance of its new 787 Dreamliner aircraft for airline operators.[28] The supply of remanufactured parts and remanufacturing services is closely correlated with the availability of skilled and certified maintenance and repair technicians. MROs constantly recruit qualified workers and seek to train new workers when possible.[29] However, because the skills needed to provide MRO services and repair parts are highly technical and require much experience and training, the pool of qualified workers is relatively limited.

Estimates of U.S. Trade in Remanufactured Aerospace Products and Cores

Trade in Remanufactured Aerospace Products

Exports of Remanufactured Aerospace Products

Between 2009 and 2011, annual U.S. exports of remanufactured aerospace products are estimated to have remained steady at about $2.5 billion (table 3.3). U.S. exports of remanufactured aerospace products are estimated to have accounted for 28–30 percent of total U.S. exports of aerospace products (new and remanufactured) during the period ($8.5 billion in 2011).[30] Markets in the EU, Canada, and Japan collectively accounted for 25 percent of U.S. exports of remanufactured aerospace products during the period, reflecting well-developed and mature aircraft maintenance industries, either through maintenance and repair facilities or at the airlines in these markets.

Between 2009 and 2011, U.S. exports to FTA partners are estimated to have accounted for approximately 15 percent ($369 million) of total U.S. exports of remanufactured aerospace products (table 3.4). Canada alone is estimated to have accounted for 40 percent

($148 million) of U.S. exports to FTA partners in 2011, as the country possesses a well-developed aircraft maintenance industry.

Table 3.3. Aerospace: U.S. exports of remanufactured goods by leading destination, 2009–11

Country or market	2009	2010	2011	% change, 2009–11
		Thousand $		
EU	477,623	409,344	395,011	–17.3
Canada	137,259	164,292	147,574	7.5
Japan	148,885	82,279	96,714	–35.0
All other countries or markets	1,782,812	1,869,668	1,950,244	9.4
Total	2,546,579	2,525,582	2,589,543	1.7

Source: USITC staff calculations of weighted responses to the Commission questionnaire.
Note: Totals may not sum due to rounding.

Table 3.4. Aerospace: U.S. exports of remanufactured goods to FTA partners, 2009–11

FTA partner	2009	2010	2011	% change, 2009–11
		Thousand $		
Canada	137,259	164,292	147,574	7.5
Singapore	37,391	91,606	76,472	104.5
Mexico	56,448	35,634	48,358	–14.3
All other FTA partners	136,910	103,062	96,470	–29.5
Total	368,007	394,594	368,873	0.2

Source: USITC staff calculations of weighted responses to the Commission questionnaire. Note: Totals may not sum due to rounding.

Table 3.5. Aerospace: U.S. exports of remanufactured goods by foreign remanufacturers invested in the United States by leading destination, 2009–11

Country or market	2009	2010	2011	% change, 2009–11
		Thousand $		
EU	118,293	110,531	97,331	–17.7
All other countries or markets	132,495	200,071	156,071	17.8
Total	250,788	310,602	253,402	1.0

Source: USITC staff calculations of weighted responses to the Commission questionnaire.
Note: Totals may not sum due to rounding. Figures for U.S. exports to other leading destinations were suppressed to prevent the disclosure of CBI and were included in "all other countries or markets."

In 2011, U.S. exports by foreign remanufacturers invested in the United States are estimated to have accounted for approximately 10 percent ($253 million) of total U.S. exports of remanufactured aerospace products (table 3.5). The EU is estimated to have received almost 40 percent ($97 million) of U.S. exports by foreign remanufacturers invested in the United States in 2011, and likely reflects intra-firm shipments of EU companies invested in the United States.

Imports of Remanufactured Aerospace Products

As noted in table 3.2, U.S. imports of remanufactured aerospace goods are estimated to have grown by 286 percent from $484 million in 2009 to nearly $1.9 billion in 2011. Even with this sharp increase, U.S. imports of remanufactured aerospace products are estimated to have accounted for only 2–6 percent of total U.S. imports of aerospace products (new and remanufactured) during the period ($35.4 billion in 2011).[31] U.S. imports by foreign remanufacturers invested in the United States are estimated to have accounted for 15 percent ($282 million) of total U.S. imports of remanufactured aerospace products in 2011 (figure 3.2). While U.S. imports of remanufactured aerospace products from the EU in 2011 ($316 million) made up only an estimated 17 percent of total U.S. imports of remanufactured goods in 2011, they are estimated to have accounted for 75 percent ($212 million) of total U.S. imports by foreign remanufacturers invested in the United States, reflecting European aircraft maintenance facilities located in the United States. The increase in U.S. imports of remanufactured aerospace products likely reflects the demand for aircraft maintenance owing to the age and heavy usage of the U.S. aircraft fleet.

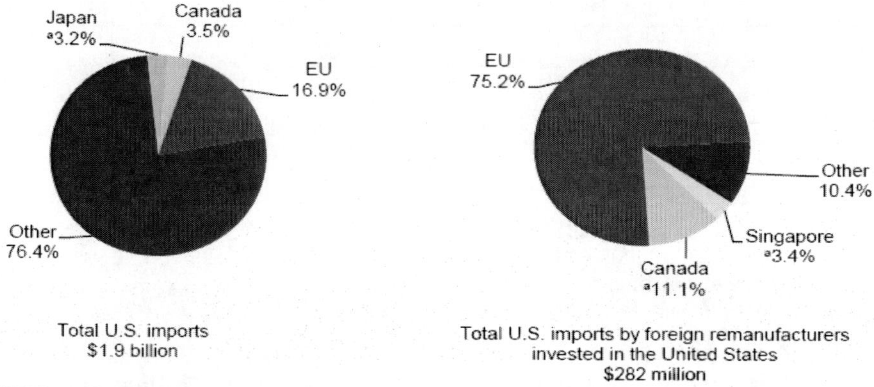

Source: USITC staff calculations of weighted responses to the Commission questionnaire.
[a]Low-precision estimate; RSE below 65 percent.
[b]Low-precision estimate; RSE below 80 percent.

Figure 3.2. Aerospace: U.S. imports of remanufactured goods by leading source, 2011.

Trade in Aerospace Cores

Exports of Aerospace Cores

Between 2009 and 2011, U.S. exports of aerospace cores are estimated to have increased by 57 percent from $383 million in 2009 to $600 million in 2011 (figure 3.3). Non-FTA partner countries are estimated to have been the destination for at least 39 percent of U.S. exports of aerospace cores in 2011 (table 3.4), with most cores likely going to the EU. U.S. exports of aerospace cores by foreign remanufacturers invested in the United States are estimated to account for 7 percent ($39 million) of total U.S. exports of aerospace cores in 2011; again, most cores were likely destined to the EU.

Imports of Aerospace Cores

Between 2009 and 2011, U.S. imports of aerospace cores are estimated to have more than doubled, rising from $43 million in 2009 to $94 million in 2011. Most U.S. imports are estimated to come from non-FTA partner countries (figure 3.5), with the majority likely coming from the EU.

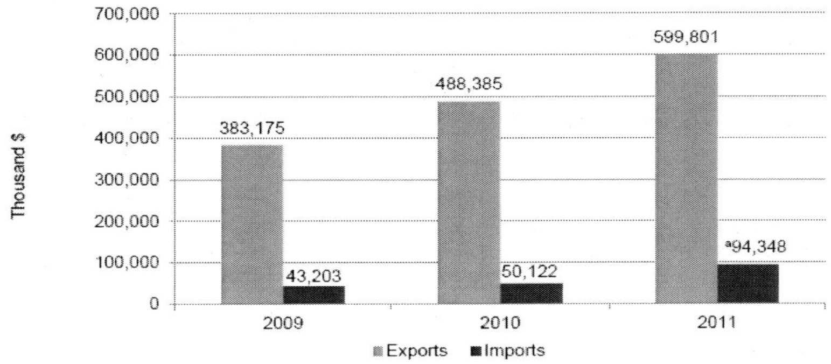

Source: USITC staff calculations of weighted responses to the Commission questionnaire.
[a]Low-precision estimate; RSE below 65 percent.

Figure 3.3. Aerospace: U.S. core exports and imports, 2009–11.

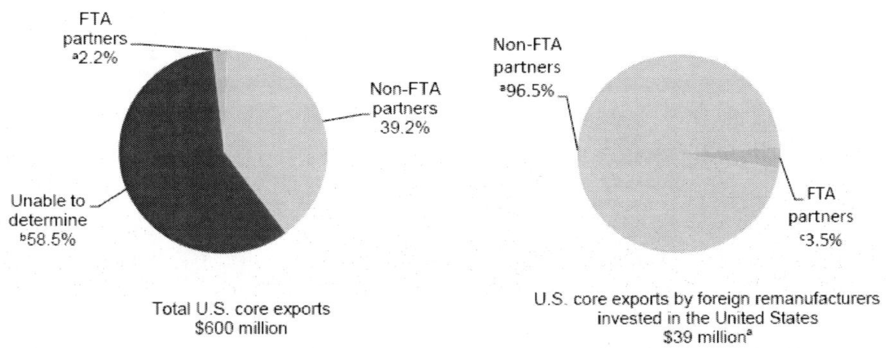

Source: USITC staff calculations of weighted responses to the Commission questionnaire.

Note: USITC staff was unable to estimate U.S. exports of cores by leading destination because of the low statistical significance of the weighted responses to the Commission questionnaire. Instead, USITC staff estimated U.S. exports of cores to FTA and non-FTA partners as two separate groups, which had higher statistical significance.

[a]Low-precision estimate; RSE below 65 percent.

[b]Questionnaire respondents were asked to report the value and destination of their core exports. In some instances, respondents reported the value, but not the destination. In these cases, USITC staff was unable to determine the destination of U.S. core exports.

[c]Low-precision estimate; RSE below 100 percent.

Figure 3.4. Aerospace: U.S. core exports to FTA versus non-FTA partners, 2011.

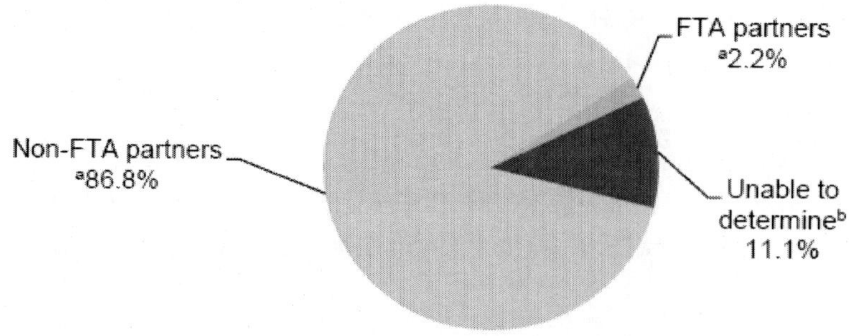

Source: USITC staff calculations of weighted responses to the Commission questionnaire.

Note: USITC staff was unable to estimate U.S. imports of cores from leading source because of the low statistical significance of the weighted responses to the Commission questionnaire. Instead, USITC staff estimated U.S. imports of cores from FTA and non-FTA partners as two separate groups, which had higher statistical significance.

[a]Low-precision estimate; RSE below 80 percent.

[b]Questionnaire respondents were asked to report the value and source of their core imports. In some instances, respondents reported the value, but not the source. In these cases, USITC staff was unable to determine the source of U.S. core imports.

[c]Low-precision estimate; RSE below 65 percent.

Figure 3.5. Aerospace: U.S. core imports from FTA versus non-FTA partners, 2011.

Barriers to Trade in Remanufactured Aerospace Products

Few significant trade barriers seem to hinder the efficient operation of the global MRO industry. The exceptions are certification requirements and foreign recognition of domestic certification standards. For instance, repair stations reportedly have difficulties obtaining certification to perform repairs in other countries.[32]

FDI in the Aerospace Remanufacturing Sector

Between 2009 and 2011, U.S. direct investment flows in foreign aerospace remanufacturing activities are estimated to have more than quadrupled, from $11.5 million in 2009 to $47.7 million in 2011 (table 3.6). Leading maintenance and repair providers (OEMs, airline operators and their maintenance subsidiaries, MROs, and independent suppliers) tend to have a global presence and are therefore invested in other regions besides their home country. For example, major U.S.-headquartered OEMs such as General Electric have service centers and subsidiaries in regions other than the United States. Similarly, major non-U.S.-based suppliers, such as EU-based Lufthansa Technik, have service centers and subsidiaries in regions other than the EU, including in the United States.

During the same period, FDI flows in U.S. aerospace remanufacturing activities are estimated to have grown by 22 percent from $34 million in 2009 to $41 million in 2011 (table 3.6); the majority came from the EU. U.S. remanufacturing employment in the aerospace sector that is associated with FDI in U.S. remanufacturing is estimated to have increased modestly from about 4,400 workers in 2009 to 4,900 workers in 2011.

Table 3.6. Aerospace: FDI and associated employment in U.S. remanufacturing, 2009–11

Item	2009	2010	2011	% change, 2009–11
	Thousand $			
Outbound FDI	11,468	70,758	47,708	316.0
Inbound FDI	[a]33,699	[a]27,193	[a]41,051	21.8
	Full-time workers			
U.S. remanufacturing employment associated with inbound FDI	4,403	4,648	4,872	10.7

Source: USITC staff calculations of weighted responses to the Commission questionnaire.
Note: Totals may not sum due to rounding.
[a]Low-precision estimate; RSE below 65 percent.

REFERENCES

AeroStrategy Management Consulting. *Global MRO Market Economic Assessment.* Prepared for the Aeronautical Repair Station Association (ARSA). Ann Arbor, MI, August 21, 2009. http://www.arsa.org/files /ARSACivilAircraftMROMarketOverview-20090821.pdf.

Airbus, SAS. *Global Market Forecast, 2010–30.* Blagnac, France, 2011.

Airlfleets.net. “Airline Fleet Age.” http://www.airfleets.net/ageflotte/fleet-age.htm (accessed June 18, 2012).

Ang, Chye Kiat. Panelist, “What’s Hot: Special Report; MRO Industry Leaders Discuss the Top Current Issues.” Aviation Week MRO Americas Conference, Dallas, TX, April 5, 2012.

Boeing. “787 GoldCare Airplane Business Solutions.” http://www.boeing.com /commercial/goldcare/index.html (accessed May 1, 2012).

———. “Boeing Edge Sets New Standard for Aviation Services and Support.” Press release, February 15, 2012.

———. *Current Market Outlook, 2010–30.* Seattle, WA: Boeing, 2011.

Butler, Phil. Product Group Report: Aerospace; An Assessment of the Remanufacturing of Aerospace Components and Aircraft. Centre for Remanufacturing & Reuse (CRR), May 2009.

Doan, Chris. “The Global MRO Forecast 2012–2022.” PowerPoint presentation, Aviation Week MRO Americas Conference, Dallas, Texas, April 3, 2012.

Michaels, Kevin. “Seven Paradigm Shifts Reshaping Air Transport MRO.” PowerPoint presentation, Aviation Week MRO Americas Conference, Dallas, TX, April 3, 2012.

Swearingin, Mark. Panelist, “Customers Speak Out.” Aviation Week MRO Americas Conference, Dallas, TX, April 4, 2012.

U.S. Department of Transportation (USDOT), Federal Aviation Administration (FAA). “Forecast of Top 10 Traffic Flows.” Washington, DC, March 9, 2012.

HEAVY-DUTY AND OFF-ROAD (HDOR) EQUIPMENT

Description of HDOR Remanufacturing Activities

The HDOR equipment remanufacturing sector includes firms that remanufacture components or parts of off-highway equipment generally used in the construction, farming, mining, and oil and gas drilling industries. Examples of HDOR equipment include trucks, bulldozers, excavators, backhoes, asphalt pavers and rollers, farm tractors and combines, rock cutters and tunneling machinery, and oil and gas drilling machinery. Remanufactured components and parts typically include diesel engines and engine components, starters, alternators, turbochargers, fuel pumps and nozzles, transmission axles, and hydraulic and electronic components.

The process for remanufacturing HDOR equipment typically involves disassembly, cleaning, testing, and repair activities. Each component is thoroughly cleaned and inspected for wear and damage. Worn, missing, or damaged components are then replaced with new or restored components that incorporate engineering updates where possible. Electrical parts are rewound or rewired, as necessary. Components are then reassembled and tested to ensure performance to original working condition or better. More than 70 percent of HDOR equipment remanufacturers are estimated to remanufacture products to better-than-original working condition.[1]

Size of HDOR Remanufacturing in the United States

Industry Structure

An estimated 200–300 firms remanufacture HDOR equipment in the United States, ranging in size from small-job shops employing fewer than 10 employees to large multinational firms employing thousands of workers.[2] HDOR remanufacturers include OEMs that produce new HDOR equipment, licensed distributors of OEMs, independent remanufacturers, contract remanufacturers that perform remanufacturing services for customers, and smaller-job shops that perform remanufacturing services for customers on a walk-in basis. U.S.-based multinational OEMs such as Caterpillar, John Deere, and Cummins are among the largest HDOR equipment remanufacturers in the United States, and produce both new and remanufactured HDOR equipment in other countries, including China, Brazil, and India. Other large HDOR remanufacturers located in the United States include Remy International and Detroit Diesel, remanufacturers of both HDOR equipment and motor vehicle parts. Many HDOR producers recognize the value of remanufacturing, and produce and/or sell remanufactured HDOR equipment along with new products as a part of their aftermarket product offerings. An estimated one-third of the firms in the HDOR sector produce both new and remanufactured goods. In contrast, an estimated 95 percent of firms sell new HDOR equipment in addition to remanufactured HDOR equipment, reflecting a desire to provide a more diverse aftermarket product offering to customers. In addition, adding remanufactured goods to existing new product offerings reportedly can help firms reduce their new-product inventories and shorten lead times for customers. [3] OEMs tend to remanufacture a range of their own new products, while other HDOR remanufacturers may focus on remanufacturing only a few products, such as alternators or starters.[4]

Production, Investment, and Employment

The United States is estimated to be the world's largest producer of remanufactured HDOR equipment. Between 2009 and 2011, U.S. production of remanufactured HDOR equipment is estimated to have increased by 50 percent from $5.2 billion in 2009 to $7.8 billion in 2011 (table 4.1). In spite of this growth, remanufactured HDOR equipment is estimated to have accounted for only 3–4 percent of total sales by U.S. firms in the HDOR equipment sector during the period ($205 billion in 2011).[5]

Table 4.1. HDOR: Remanufacturing production, investment, and employment, 2009–11

Item	2009	2010	2011	% change, 2009–11
		Thousand $		
Production	5,152,938	6,244,302	7,770,586	50.8
Investment	[a]80,183	[a]131,963	[a]162,746	103.0
		Full-time workers		
Employment	18,072	18,654	20,870	15.5

Source: USITC staff calculations of weighted responses to the Commission questionnaire. [a]Low-precision estimate; RSE below 65 percent.

Between 2009 and 2011, HDOR remanufacturing investment is estimated to have doubled to $163 million, although the effect was far from uniform: only an estimated one-third of U.S. HDOR equipment remanufacturers increased production capacity during the period. At the same time, U.S. remanufacturing employment in the HDOR equipment sector is estimated to have increased by about 15 percent to approximately 21,000 workers.

Factors Affecting Trends in Sales and Investment

Primary factors affecting recent trends in sales of remanufactured HDOR equipment include the lack of a common definition of remanufactured goods, competition with lower-cost imports of new products, and advances in manufacturing processes for new products. In the U.S market, imports of lower-cost new HDOR parts reportedly compete directly with domestic remanufactured HDOR equipment, such as starters and alternators. Presently, some domestic OEM HDOR equipment remanufacturers report having a cost advantage over imports of low-priced "knock off" new products. They also tout the higher quality and reliability of their products, as well as their ability to provide customer support through their dealer networks.[6] Another challenge facing HDOR equipment remanufacturers involves changes or advances in manufacturing processes for new products, which have reportedly required increased investment in and changes to remanufacturing processes. For instance, the manufacturing process for some new components has increasingly focused on reducing component weight. As a result, some new components that were previously designed and produced with excess materials to be machined away are no longer produced this way. Instead, remanufacturers have reportedly developed unique advanced technologies, such as lighter-weight spray welding and laser cladding, that add less metal to worn parts. In other instances, OEM remanufacturers may design new products with remanufacturing in mind, leading to more collaborative R&D and engineering efforts between manufacturing and remanufacturing teams.[7] Other factors cited by HDOR equipment remanufacturers that affect their competitive position include high transportation costs (cited by 36 percent of U.S. HDOR equipment remanufacturers), environmental regulations (31 percent), licensing or certification requirements (26 percent), and healthcare costs (24 percent).[8]

U.S. Market for Remanufactured HDOR Equipment

Market Size

The U.S. market for remanufactured HDOR equipment is estimated to have increased by more than 50 percent from $4.5 billion in 2009 to $6.8 billion 2011 (table 4.2). About 30 percent of annual U.S. production of remanufactured HDOR equipment was exported in 2011, while imports of remanufactured HDOR equipment are estimated to have accounted for around 22 percent of U.S. apparent consumption during the period. One source has estimated the size of the U.S. market for remanufactured HDOR construction equipment (such as backhoe loaders and excavators) and agriculture equipment (such as farm tractors and combines)—two segments of the broader remanufactured HDOR equipment market—to be $1.4 billion and $1.1 billion, respectively, in 2011. [9] Both segments held a substantial share of their respective aftermarket segments. For instance, remanufactured HDOR construction equipment accounted for an estimated 18 percent ($1.4 billion) of the $7.4 billion HDOR construction equipment aftermarket (new and remanufactured), while remanufactured HDOR agriculture equipment accounted for an estimated 21 percent ($1.1 billion) of the $5.3 billion HDOR agriculture equipment aftermarket.

Table 4.2. HDOR: U.S. market (apparent consumption) for remanufactured goods, 2009–11

Item	2009	2010	2011	% change, 2009–11
		Thousand $		
Production	5,152,938	6,244,302	7,770,586	50.8
Imports	1,042,284	1,302,481	1,489,259	42.9
Exports	[a]1,695,950	[a]2,045,076	[a]2,451,967	44.6
U.S. apparent consumption	4,499,272	5,501,707	6,807,878	51.3

Source: USITC staff calculations of weighted responses to the Commission questionnaire. Note: Totals may not sum due to rounding.

[a]Low-precision estimate; RSE below 65 percent.

U.S. Buyers and Demand Factors

Buyers of remanufactured HDOR equipment include construction companies and independent construction operators, mining companies, energy exploration companies, and commercial and independent farmers. Demand for HDOR equipment (either new or remanufactured) is thus influenced by the level of current and projected construction, mining, and agriculture activities at a given time. Factors that affect the demand for remanufactured HDOR equipment primarily include price and customer perceptions of the quality and value of the product. Remanufactured HDOR equipment may cost consumers up to 70 percent less than comparable new products. [10] Some industry representatives assert that demand for remanufactured HDOR equipment may be somewhat countercyclical, increasing during periods of economic uncertainty as consumers seek to economize by switching from purchasing new products to remanufactured ones.[11] However, other industry representatives report that demand is stable regardless of the strength of the overall economy.[12] Customer perceptions of the quality and value of remanufactured HDOR equipment may also influence demand for the products. For instance, some customers may be uncertain as to the quality of remanufactured HDOR equipment, and thus hesitate to purchase it.[13] To allay concerns, HDOR equipment remanufacturers may offer various types of warranties on their products, including coverage for a longer period of time for their remanufactured parts than for equivalent new parts.

U.S. Sellers and Supply Factors

The principal sellers of remanufactured HDOR equipment include OEMs and their licensed distributors or service centers, independent remanufacturers, and smaller-job shops. Core availability, quality, and costs are the primary factors affecting the supply of remanufactured HDOR equipment. More than one-quarter of HDOR equipment remanufacturers identify scarcity and the high price of cores as "extremely important" factors affecting their ability to compete in the U.S. market. HDOR equipment remanufacturers procure cores through warranty returns, trade-ins from customers and distributors, and owner returns, all typically on some form of exchange basis. Core collection processes vary among HDOR equipment remanufacturers. The quality of the core typically determines its price and the amount of core refunds HDOR remanufacturers provide to customers. For instance, HDOR equipment remanufacturers typically provide full core refunds to customers for better quality cores, but only partial refunds for damaged ones.[14] The availability of skilled labor influences the supply of remanufactured goods. Skilled technicians and engineers are needed to detect failures that may be difficult to find during the disassembly and inspection of cores.[15] Many of the larger HDOR remanufacturers have reportedly made substantial investments in hiring and/or training skilled labor. Although some industry representatives indicate that they are encouraged by the growth in remanufacturing and potential for increased employment, an estimated 14 percent of HDOR remanufacturers consider the scarcity of skilled workers as an extremely important competitive challenge.[16] To address this scarcity, some HDOR remanufacturers have sought to hire employees with a basic level of education and math skills, and then augment these skill sets with in-house training, government-sponsored training, and education programs and internships provided in partnership with local technical colleges.[17]

Estimates of U.S. Trade in Remanufactured HDOR Equipment and Cores

Trade in Remanufactured HDOR Equipment

Exports of Remanufactured HDOR Equipment

The United States is a net exporter of remanufactured HDOR equipment. Between 2009 and 2011, U.S. exports of remanufactured HDOR equipment are estimated to have increased by 45 percent from $1.7 billion to $2.5 billion (table 4.3). At the same time, U.S. exports of remanufactured HDOR equipment are estimated to have accounted for only 5–6 percent of total U.S. exports of all HDOR equipment (new and remanufactured) during the period ($43.9 billion in 2011).[18] Canada, Australia, and Mexico were the largest markets for those U.S. exports. Between 2009 and 2011, U.S. exports of remanufactured HDOR equipment to FTA partners accounted for approximately one-half of total U.S. exports of remanufactured HDOR equipment and are estimated to have increased by almost 50 percent, growing from $889 million in 2009 to $1.3 billion in 2011 (table 4.4). U.S. exports of remanufactured HDOR equipment by foreign remanufacturers invested in the United States are estimated to account for about 1 percent ($26 million) of total U.S. exports in 2011 (table 4.5), reflecting the rarity of foreign HDOR remanufacturers invested in U.S. remanufacturing activities.

Table 4.3. HDOR: U.S. exports of remanufactured goods by leading destination, 2009–11

Country	2009	2010	2011	% change, 2009–11
		Thousand $		
Canada	359,942	455,064	561,680	56.0
Australia	[a]161,969	[a]233,338	[a]297,180	83.5
Mexico	[a]166,381	152,753	213,569	28.4
All other countries	[a]1,007,658	[a]1,203,921	[a]1,379,538	36.9
Total	[a]1,695,950	[a]2,045,076	[a]2,451,967	44.6

Source: USITC staff calculations of weighted responses to the Commission questionnaire. Note: Totals may not sum due to rounding.

[a]Low-precision estimate; RSE below 65 percent.

Table 4.4. HDOR: U.S. exports of remanufactured goods to FTA partners, 2009–11

FTA partner	2009	2010	2011	% change, 2009–11
		Thousand $		
Canada	359,942	455,064	561,680	56.0
Australia	[a]161,969	[a]233,338	[a]297,180	83.5
Mexico	[a]166,381	152,753	213,569	28.4
All other FTA partners	[a]200,855	[a]239,322	[a]243,378	21.2
Total	889,147	1,080,477	1,315,807	48.0

Source: USITC staff calculations of weighted responses to the Commission questionnaire. Note: Totals may not sum due to rounding.

[a]Low-precision estimate; RSE below 65 percent.

Table 4.5. HDOR: U.S. exports of remanufactured goods by foreign remanufacturers invested in the United States by leading destination, 2009–11

Country	2009	2010	2011	% change, 2009–11
		Thousand $		
Canada	([a])	([a])	([a])	([a])
Mexico	([a])	([a])	([a])	([a])
Chile	([a])	([a])	([a])	([a])
All other countries	[b]229	[c]1,802	1,710	646.7
Total	[b]8,204	[b]15,902	[c]26,364	221.4

Source: USITC staff calculations of weighted responses to the Commission questionnaire. Note: Totals may not sum due to rounding.

[a]Figures suppressed to prevent the disclosure of CBI. [b]Low-precision estimate; RSE below 65 percent. [c]Low-precision estimate; RSE below 80 percent.

Table 4.6. HDOR: U.S. imports of remanufactured goods by leading source, 2011

Country or market	2011	% share
	Thousand $	
Mexico	[a]1,325,206	89.0
Canada	[b]78,941	5.3
EU	[b]35,544	2.4
All other countries or markets	49,568	3.3
Total	1,489,259	100.0

Source: USITC staff calculations of weighted responses to the Commission questionnaire.

Note: Figures for U.S. imports from "all other countries or markets" are calculated. Totals may not sum due to rounding.

[a]Low-precision estimate; RSE below 65 percent. [b]Low-precision estimate; RSE below 80 percent.

Imports of Remanufactured HDOR Equipment

As noted in table 4.2, U.S. imports of remanufactured HDOR equipment are estimated to have increased by 43 percent to $1.5 billion between 2009 and 2011. At the same time, U.S. imports of remanufactured HDOR equipment are estimated to have accounted for 6–7 percent of total U.S. imports of all HDOR equipment (new and remanufactured) during the same period ($27.1 billion in 2011).[19] Mexico is estimated to have accounted for almost 90 percent of U.S. imports in 2011 (table 4.6), reflecting cross-border production networks in which cores are exported to Mexico to be remanufactured and then reimported for sale in the U.S. market. Examples reportedly include several large HDOR equipment and motor vehicle parts producers, including Caterpillar, Cummins, and Remy. [20] Figures for U.S. imports of remanufactured HDOR equipment by foreign remanufacturers invested in the United States are not available.[21]

Trade in HDOR Cores

Exports of HDOR Cores

The United States is a net importer of HDOR equipment cores (figure 4.1). However, between 2009 and 2011, U.S. exports of HDOR cores are estimated to have increased modestly, from $240 million in 2009 to $256 million in 2011. The vast majority of these cores were destined for FTA partners, most likely Mexico or Canada (table 4.7). Figures for U.S. exports of HDOR cores by foreign remanufacturers invested in the United States are not available.[22]

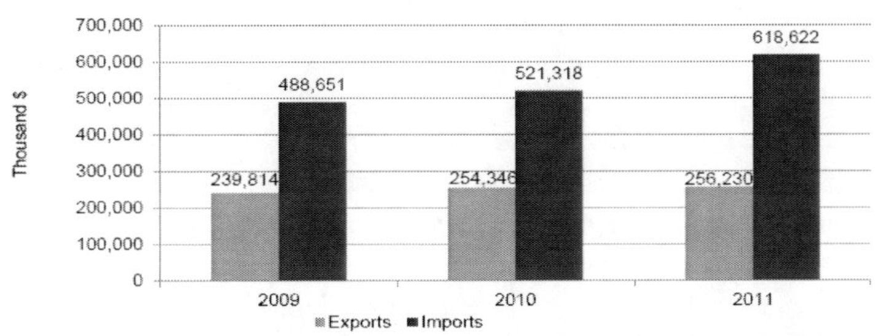

Source: USITC staff calculations of weighted responses to the Commission questionnaire.

Figure 4.1. HDOR: U.S. core exports and imports, 2009–11.

Imports of HDOR Cores

During this same period, U.S. imports of cores of HDOR equipment are estimated to have increased by 27 percent from $489 million in 2009 to $619 million in 2011. U.S. imports from FTA partner countries are estimated to have accounted for at least one-third of total U.S. imports of HDOR cores in that year (table 4.8). Figures for U.S. imports of HDOR cores by foreign remanufacturers invested in the United States are not available.[23]

Table 4.7. HDOR: U.S. core exports to FTA versus non-FTA partners, 2011

Export market	2011 Thousand $	% share
FTA partners	249,252	97.3
Non-FTA partners	4,630	1.8
Unable to determine[a]	2,348	0.9
Total	256,230	100.0

Source: USITC staff calculations of weighted responses to the Commission questionnaire.

Note: USITC staff was unable to estimate U.S. exports of cores by leading destinations because of the low statistical significance of the weighted responses to the Commission questionnaire. Instead, USITC staff estimated U.S. exports of cores to FTA and non-FTA partners as two separate groups, which had higher statistical significance. Totals may not sum due to rounding.

[a]Questionnaire respondents were asked to report the value and destination of their core exports. In some instances, respondents reported the value, but not the destination. In these cases, USITC staff was unable to determine the destination of U.S. core exports.

Table 4.8. HDOR: U.S. core imports from FTA versus non-FTA partners, 2011

Leading source	2011 Thousand $	% share
FTA partners	207,822	33.6
Non-FTA partners	[a]123,952	20.0
Unable to determine[b]	286,849	46.4
Total	618,622	100.0

Source: USITC staff calculations of weighted responses to the Commission questionnaire.

Note: USITC staff was unable to estimate U.S. imports of cores by leading sources because of the low statistical significance of the weighted responses to the Commission questionnaire. Instead, USITC staff estimated U.S. imports of cores from FTA and non-FTA partners as two separate groups, which had higher levels of statistical significance. Totals may not sum due to rounding.

[a]Low-precision estimate; RSE below 80 percent.

[b]Questionnaire respondents were asked to report the value and source of their core imports. In some instances, respondents reported the value, but not the source. In these cases, USITC staff was unable to determine the source of U.S. core imports.

Barriers to Trade in Remanufactured HDOR Equipment and Cores

Some industry representatives reported that the single largest obstacle to increased sales in foreign markets is the lack of a common, global definition and set of standards that distinguishes remanufactured products from used products.[24] Many countries reportedly limit or ban trade in remanufactured HDOR equipment and related cores. Witnesses at the Commission hearing cited barriers to trade in remanufactured HDOR equipment and cores in

Brazil, China, Ecuador, India, Peru, Turkey, and Russia.[25] Nontariff restrictions reportedly include prohibitive taxes and registration requirements, complex paperwork requirements, onerous customs regulations and practices, and domestic preference laws that favor domestically produced goods over imported ones.[26]

HDOR equipment remanufacturers operating in countries that limit or prohibit core imports reportedly face a domestic shortage of cores. In these countries, remanufacturers may offer a trade-in discount on sales of new products to provide an incentive to the customers to return the core, and thus help increase the supply of domestic cores.[27] However, industry representatives report that this practice is not economically sustainable in the long run.[28]

FDI in the HDOR Equipment Remanufacturing Sector

Data on FDI in the HDOR equipment remanufacturing sector in the United States are not available, but such FDI is likely to be relatively modest. U.S. HDOR remanufacturing employment associated with FDI is estimated to account for less than 2 percent (390– 520 workers) of total U.S. HDOR remanufacturing employment during 2009–11. [29] Komatsu, a Japanese HDOR equipment producer, reportedly invested in a remanufacturing facility in Lexington, KY, but later sold the plant to Springfield Remanufacturing Center (Springfield) in September 2009. Springfield is now a supplier of remanufactured equipment to Komatsu.[30]

By contrast, between 2009 and 2011, U.S. direct investment flows in foreign HDOR remanufacturing activities are estimated to have tripled, jumping from $37 million in 2009 to $111 million in 2011 (table 4.9). The EU received an estimated 36 percent ($40 million) of U.S. direct investment in 2011, while Mexico received an estimated 29 percent ($32 million). Some of the largest U.S.-based HDOR equipment remanufacturers, including Caterpillar, Cummins, and John Deere, have remanufacturing facilities located in or near foreign markets experiencing growth in demand for aftermarket parts. [31] For instance, Caterpillar reportedly locates facilities where the company can move cores freely and where it can leverage its operating scale within a three-market radius.[32] In May 2011, Caterpillar opened a remanufacturing facility in Singapore as its regional source for remanufactured components to serve Asia's mining industry.[33]

Table 4.9. HDOR: U.S. outbound FDI in foreign remanufacturing by destination, 2009–11

Country or market	2009	2010	2011	% change, 2009–11
		Thousand $		
EU	([a])	[b]8,149	[b]39,919	n/a
Mexico	10,820	21,539	32,413	199.6
All other countries or markets	25,686	70,031	38,935	51.6
Total	36,505	99,719	111,267	204.8

Source: USITC staff calculations of weighted responses to the Commission questionnaire. Note: Totals may not sum due to rounding. N/a = not applicable.

[a]None reported. [b]Low-precision estimate; RSE below 80 percent.

References

Bosch Rexroth, Corp. "Reman Exchange Program," n.d.
http://www.boschrexroth.com/spc/SVH/REMAN/REMAN.cfm?cfid=7535335&cftoken=972 343 86&Country=default&Languag (accessed August 26, 2011).

Caterpillar Inc. "Cat Remanufactured 3054 Engine for Backhoe Loader Applications: Core Acceptance Criteria for 3054 Engines," 2004. http://www.cat.com/cda/files /280395/9/pehj0050.pdf.

———. "Caterpillar Expands Sustainable Product Offerings in Asia-Pacific with Opening of New Remanufacturing Facility in Singapore." Press release, May 24, 2011. http://catreman.cat.com/cda/components/full ArticleNoNav?x=7&id=1182323&mode.

Diesel & Gas Turbine Worldwide. "Remanufacturing versus Rebuilding," March 2012.

Muzinic, Jennifer. "SRC Integrates Newly Acquired Lexington Plant." *Springfield Business Journal Online*, May 3, 2010. http://sbj.net/Main. asp?SectionID=18&SubSectionID= 23&ArticleID=86879.

U.S. International Trade Commission (USITC). Hearing transcript in connection with inv. no. 332-525, *Remanufactured Goods: An Overview of the U.S. and Global Industries, Markets, and Trade*, February 28, 2012.

Motor Vehicle Parts[1]

Description of Motor Vehicle Parts Remanufacturing Activities

The motor vehicle parts remanufacturing industry is reportedly the world's largest remanufacturing sector, accounting for an estimated two-thirds of global remanufacturing activities. This sector includes firms that remanufacture components for passenger vehicles, such as passenger cars and light trucks, and for commercial vehicles, such as medium and heavy trucks. The remanufacturing sector is part of the larger aftermarket for motor vehicle goods.[2] The principal remanufactured products in the motor vehicle sector are engines and transmissions, starter motors and alternators, steering racks, and clutches. Remanufactured motor vehicle parts cost on average 20–50 percent less than most new parts.[3]

Compared with manufacturing new motor vehicle parts, remanufacturing tends to be more labor intensive, with numerous disassembly, separation, cleaning, and repair tasks.[4] A remanufacturing facility reportedly requires three times as many employees as a typical manufacturing operation.[5] On the other hand, remanufacturing requires more limited investment in machinery and equipment.[6]

The component to be remanufactured (the core) must be completely disassembled, cleaned, and examined for wear and breakage. Worn, missing, or nonfunctioning components are replaced with new or remanufactured parts, while electrical parts are rewound or rewired, as necessary. After these processes are completed, the part is reassembled and tested for compliance with existing performance specifications.[7]

An estimated 84 percent of U.S. producers that remanufacture motor vehicle parts indicated that their remanufactured product exceeds the original working condition of the good.[8]

Size of Motor Vehicle Parts Remanufacturing in the United States

Industry Structure

The United States is the headquarters for many of the world's leading motor vehicle parts remanufacturers and is home to many other smaller remanufacturers. Estimates of industry size vary widely, depending in part on the scope of remanufacturing activities included in the estimate. There are an estimated 300 U.S. remanufacturers of motor vehicle parts that employ more than 20 workers each, although thousands of smaller companies (with fewer than 20 workers per firm) reportedly perform some steps of the remanufacturing process on a range of motor vehicle parts. [9] According to the Rochester Institute of Technology (RIT), roughly 3,800 U.S. firms remanufacture motor vehicle parts, whereas the Automotive Parts Remanufacturers Association (APRA) identifies more than 15,000 U.S. firms as remanufacturers of motor vehicle parts.[10] Firms such as Remy International, Cardone Industries, Caterpillar, Cummins, TRW, and Meritor are among the leading motor vehicle parts and HDOR equipment remanufacturing firms headquartered in the United States, employing hundreds of workers at U.S. and foreign locations.[11] Many other remanufacturers specialize in remanufacturing one or two types of motor vehicle components, such as alternators and starters.[12]

Remanufacturers of motor vehicle parts include original equipment manufacturers (OEMs, or in this case, the vehicle makers),[13] original equipment suppliers (OESs), and independent remanufacturing firms. OEMs produce new motor vehicle parts and also remanufacture or contract out the remanufacturing of their products for resale. OEMs typically sell their remanufactured products through their dealer networks. [14] OES remanufacturers produce new motor vehicle parts and also remanufacture their products to supply the automakers. Independent remanufacturing firms, on the other hand, have limited or no contractual relationships with OEMs or OESs, and tend to remanufacture and sell their goods largely outside of the OEM and OES distribution channels.[15]

Another distinction between OEMs/OESs and independent remanufacturers is the way in which the firms design the remanufactured goods. Because they design and produce original equipment (OE) motor vehicle parts, OEM and OES remanufacturers are able to produce remanufactured goods to the original equipment specifications of the motor vehicle parts, based on original design and product parameters. Independent remanufacturers, on the other hand, must reverse-engineer the motor vehicle parts they remanufacture, since they lack the original equipment designs and specifications. Using this reverse-engineering process, independent remanufacturers develop a design and establish internal product specifications that may or may not match those of the original equipment components, but still meet form, fit, and function requirements.[16]

Production, Investment, and Employment

The motor vehicle parts remanufacturing industry is the third largest remanufacturing sector in the United States, with domestic production estimated at $6.2 billion in 2011, down from an estimated $7.0 billion in 2009 (table 5.1). Remanufactured motor vehicle parts are estimated to have accounted for 1–2 percent of total sales by U.S. firms within the motor vehicle parts sector during the study period ($553 billion in 2011).[17] The decrease in production during the period may reflect declining demand for certain remanufactured motor vehicle parts. For instance, some vehicle fleet customers (major consumers of remanufactured motor vehicle parts) reportedly delayed purchases of remanufactured powertrains in an effort to save money. [18] Despite this decline, an estimated 46 percent of firms that remanufacture motor vehicle parts experienced no change in production capacity during the period. Another 30 percent increased production capacity, whereas only 19 percent of remanufacturers reduced production capacity.

Table 5.1. Motor vehicle parts: Remanufacturing production, investment, and employment, 2009–11

Item	2009	2010	2011	% change, 2009–11
	Thousand $			
Production	7,018,303	6,969,823	6,211,838	-11.5
Investment	76,106	105,669	105,684	38.9
	Full-time workers			
Employment	30,069	30,404	30,653	1.9

Source: USITC staff calculations of weighted responses to the Commission questionnaire.

Total remanufacturing investment is estimated to have increased by about 50 percent from $76 million in 2009 to $106 million in 2011. Meritor, for example, invested $2.7 million in equipment for its Plainfield, IN, plant, which remanufactures commercial truck components.[19] The increased presence of foreign automakers in the U.S. market has also attracted investment from their network of parts manufacturers that produce both new and remanufactured goods. [20] Annual U.S. employment in motor vehicle parts remanufacturing is estimated to have ranged from 30,000 to about 31,000 full-time employees during the period.

Factors Affecting Trends in Sales, Investment, and Employment

Labor costs, transportation costs, and scarcity of cores were the top factors cited by U.S. motor vehicle parts remanufacturers that affect their ability to compete in the U.S. market.

Labor costs play an important role because of the labor-intensive nature of motor vehicle parts remanufacturing. Transportation costs and the scarcity of cores are important considerations, given the nature of core procurement and closed-loop supply chains noted in chapters 1 and 2, and further described in this chapter in the section on supply factors. Other factors cited by U.S. motor vehicle parts remanufacturers include the high price of cores, declining demand for remanufactured goods, environmental regulations, and rising competition from imported remanufactured goods.[21]

Increased availability of low-cost new goods from Asia may be reducing sales of U.S. remanufactured products and potentially reducing employment levels. For example, imports of new axles from China, Korea, and Vietnam have reportedly reduced sales of front-wheel drive axles remanufactured in the United States.[22]

U.S. Market for Remanufactured Motor Vehicle Parts

The U.S. market for remanufactured motor vehicle parts is reportedly the world's largest, and is an important subset of the U.S. market for motor vehicle parts. Numerous factors affect the demand for and supply of remanufactured motor vehicle goods in the U.S. market. The age of the U.S. motor vehicle fleet, the cost and availability of substitute products, component reliability, vehicle warranties, and the degree of consumers' cost-consciousness all affect demand.

The availability of cores and investment in remanufacturing R&D primarily affect the supply of remanufactured motor vehicle parts.

Market Size

The U.S. market for remanufactured motor vehicle parts is estimated to have dropped by 9 percent from $7.8 billion in 2009 to $7.1 billion in 2011 (table 5.2). The decline in the size of the U.S. market largely reflects a drop in U.S. production; as mentioned above, this was likely prompted by falling demand for certain remanufactured motor vehicle parts, such as powertrains. Overall, however, motor vehicle remanufacturing reportedly did not decline as much as other industries during the downturn, and some segments of the industry reportedly experienced an increase in demand for remanufactured motor vehicle parts.[23]

Table 5.2. Motor vehicle parts: U.S. market (apparent consumption) for remanufactured goods, 2009–11

Item	2009	2010	2011	% change, 2009–11
	Thousand $			
Production	7,018,303	6,969,823	6,211,838	-11.5
Imports	1,219,922	1,462,269	1,481,939	21.5
Exports	430,145	494,145	581,520	35.2
U.S. apparent consumption	7,808,080	7,937,947	7,112,257	-8.9

Source: USITC staff calculations of weighted responses to the Commission questionnaire.
Note: Totals may not sum due to rounding.

At the same time, U.S. imports of certain remanufactured motor vehicle parts are estimated to have increased by 21 percent during 2009–11. The United States is a net importer of remanufactured motor vehicle parts. While domestic remanufacturers supplied almost 80 percent of the U.S. market, they exported only about 7.5 percent of production during the period.

U.S. Buyers and Demand Factors

U.S. Buyers

Commercial fleet owners, OEMs, and private individuals are the principal purchasers of remanufactured motor vehicle parts. Fleet owners are major demand drivers, as they have large numbers of vehicles and prioritize low-cost maintenance of their vehicles over an extended period of time.

OEMs purchase remanufactured motor vehicle goods to reduce the cost of filling warranty claims, and to fulfill their obligation to provide replacement parts for older vehicles.[24]

Private individuals also purchase remanufactured motor vehicle goods, but may not be as focused on low cost over the long term. Moreover, they tend to purchase parts only in limited quantities, usually for only one or two vehicles.

Demand Factors

Demand for remanufactured motor vehicle goods is affected primarily by the average age of vehicles in the United States, availability of product substitutes, and price consciousness of consumers. The average age of the U.S. motor vehicle fleet is currently 10.8 years old, which is the oldest average ever recorded.[25] As vehicles age, they generally require more replacement parts, at least potentially stimulating demand for remanufactured parts.[26] For instance, demand for non-OEM aftermarket parts (either new or remanufactured) is low when most parts of a vehicle are still under warranty because warranty replacements are typically fulfilled by the OEM (or its dealer).[27] However, as a vehicle ages and its warranty expires, the availability of new OEM parts usually declines. When parts begin to reach the end of their useful life and need to be replaced, consumers may purchase new OES, new aftermarket, or remanufactured components. As vehicles of a given model age and leave the fleet, motor vehicle parts suppliers shift production to higher volume OE and new aftermarket components for newer vehicles in the fleet,[28] reducing the availability of new OES or aftermarket parts for older vehicles. In this case, remanufactured parts may be the only available option for replacements, as new parts may be difficult, if not impossible, to obtain.[29] Price consciousness of consumers also affects the demand for remanufactured goods. Demand for remanufactured parts may increase during periods of economic uncertainty, since remanufactured parts tend to be less expensive than comparable new parts, and consumers may switch to remanufactured parts in order to save money. During the 2008– 09 economic downturn and slow post-2009 recovery, some consumers reportedly postponed new vehicle purchases and reduced vehicle maintenance costs by purchasing remanufactured parts rather than new ones. As a result, remanufacturers reportedly experienced an increase in demand for remanufactured engines for small passenger cars, which consumers rarely purchased before the economic downturn.[30] In the case of an individual, the decision to purchase a new or remanufactured part may be driven by the age of a customer's vehicle as well as by a customer's personal economic situation. For an older vehicle, a customer will likely choose a remanufactured part, whereas a newer vehicle will likely be fitted with a new part, particularly if the part is a major system or component (e.g., an air conditioning compressor).[31] This decision may also be influenced by the length and conditions of the vehicle's warranty.

U.S. Sellers and Supply Factors

U.S. Sellers

Sellers of remanufactured motor vehicle parts include OEMs, independent remanufacturers, retail stores, and independent repair shops. OEMs sell remanufactured products through their dealerships and distribution centers. An independent remanufacturer typically sells parts outside the OEM/OES distribution channel, marketing to warehouse distributors and retail outlets that provide parts to independent repair shops and do-it-yourself (DIY) stores; to a lesser extent, these remanufacturers will also sell to the OEMs. Retail stores and independent repair shops sell remanufactured parts alongside new OEM and aftermarket products. Independent repair shops that install new aftermarket or OE-remanufactured parts purchase them through OEM-authorized dealerships or distribution centers. Retail stores and independent repair shops tend to account for a larger share of shipments from independent remanufacturers that lack strong ties to OEMs. These independent remanufacturers also tend to be the source of store-brand remanufactured parts at retail stores.[32]

Supply Factors

The cost and supply of cores, as well as investment in R&D, affect the cost and supply of remanufactured motor vehicle parts. Cores are the most important input in remanufacturing. According to a 2009 survey, cores represented up to 19 percent of the cost of a remanufactured product for most motor vehicle parts remanufacturers, although the cost was reported to be as high as 59 percent by some.[33]

The supply of usable cores available to a remanufacturer determines the number of remanufactured motor vehicle parts that a company can produce. The supply of cores varies according to the age of the component, its scrap value, and the value of the core. When a component is first introduced in a motor vehicle, few to no used cores are available to remanufacture that part. As the part ages, however, consumers replace the component and return an increasing number of cores, creating a supply of cores viable for remanufacturing. Core availability begins to decline again as the vehicle that the part is used in becomes older and is removed from service.

The quality of the core also tends to decline with the age of the part.[34] The underlying scrap value of the metals in the part to be remanufactured can also affect core availability.[35] The higher the value of the core, in absolute terms and relative to scrap value, the greater the supply of cores available to a remanufacturer.[36] When the scrap value is closer to the core value, the core will more likely be sold for scrap and eventually melted down and recycled. A leading source of cores for the motor vehicle remanufacturing sector is product returns of end-of-use vehicle parts. Through contractual arrangements, OES firms typically supply remanufactured motor vehicle parts to the OEM dealerships in exchange for cores returned by consumers to the dealership or distributor. OEM distribution networks both collect cores and distribute remanufactured OES parts. This closed-loop system reduces uncertainty in the availability of cores by building in sources of supply.[37] In contrast to these closed-loop supply chains, independent remanufacturers tend to acquire their cores from salvage yards or third-party core brokers.[38] The quantity, quality, and type of core supplied to the salvage yards and brokers may be inconsistent, which increases the uncertainty of core supply for independent remanufacturers.

However, independent remanufacturers also tend to use a core-deposit system to encourage the return of cores and to keep supply flowing.[39] Remanufacturers conduct extensive R&D to maintain remanufacturing capabilities and improve their remanufacturing processes.[40] Their ability to re-engineer and find solutions for problems with original components affects the quantity and quality of remanufactured motor vehicle parts. Both OEMs that remanufacture and independent remanufacturers research component failures and attempt to find solutions to ensure that the remanufactured part will not fail in the same way.[41] One particular area of remanufacturing that requires extensive R&D is remanufacturing electronic components or components with embedded electronics.[42] Re-engineering can be especially challenging when the software needed to test such components is unavailable to an independent remanufacturer.[43] In response, remanufacturers have increasingly hired engineers with specialized knowledge in both mechanical and electrical engineering.[44]

Estimates of U.S. Trade in Remanufactured Motor Vehicle Parts and Cores

Trade in Remanufactured Motor Vehicle Parts

Exports of Remanufactured Motor Vehicle Parts

U.S. exports of remanufactured motor vehicle parts are estimated to have increased by 35 percent from $430 million in 2009 to $582 million in 2011 (table 5.3). Nevertheless, U.S. exports of remanufactured motor vehicle parts are estimated to have accounted for only about 1 percent of total U.S. exports of motor vehicle parts (new and remanufactured) during the period ($59.1 billion in 2011).[45] NAFTA partners Canada and Mexico are estimated to have accounted for 70 percent ($409 million) of total U.S. exports of remanufactured motor vehicle parts in 2011, reflecting the regional integration of the North American motor vehicle industry. Overall, U.S. exports to FTA partners, primarily Canada and Mexico, are estimated to have increased by 65 percent from $269 million in 2009 to $445 million in 2011, accounting for about 76 percent of total U.S. exports of remanufactured motor vehicle parts in 2011 (table 5.4). U.S. exports of remanufactured motor vehicle parts by foreign remanufacturers invested in the United States (primarily to Canada) are estimated to account for 12 percent ($72 million) of total U.S. exports in 2011, up from 6 percent ($25 million) in 2009 (table 5.5).

Table 5.3. Motor vehicle parts: U.S. exports of remanufactured goods by leading destination, 2009–11

Country	2009	2010	2011	% change, 2009–11
		Thousand $		
Canada	226,213	301,458	383,758	69.6
Mexico	16,817	21,987	25,271	50.3
All other countries	[a]187,115	170,701	172,491	–7.8
Total	430,145	494,145	581,520	35.2

Source: USITC staff calculations of weighted responses to the Commission questionnaire. Note: Totals may not sum due to rounding.

[a]Low-precision estimate; RSE below 65 percent.

Table 5.4. Motor vehicle parts: U.S. exports of remanufactured goods to FTA partners, 2009–11

FTA partner	2009	2010	2011	% change, 2009–11
		Thousand $		
Canada	226,213	301,458	383,758	69.6
Mexico	16,817	21,987	25,271	50.3
All other FTA partners	[a]26,030	[a]40,894	[a]35,709	37.2
Total	269,060	364,338	444,738	65.3

Source: USITC staff calculations of weighted responses to the Commission questionnaire. Note: Totals may not sum due to rounding.

[a]Low-precision estimate; RSE below 65 percent.

Table 5.5. Motor vehicle parts: U.S. exports of remanufactured goods by foreign remanufacturers invested in the United States by leading destination, 2009–11

Country or market	2009	2010	2011	% change, 2009–11
		Thousand $		
Canada	[a]19,740	[b]51,468	[b]68,031	244.6
All other countries or markets	[a]4,892	1,837	3,715	-24.1
Total	24,632	[b]53,305	[b]71,745	191.3

Source: USITC staff calculations of weighted responses to the Commission questionnaire. *Note:* Totals may not sum due to rounding.

[a]Low-precision estimate; RSE below 65 percent. [b]Low-precision estimate; RSE below 80 percent.

Imports of Remanufactured Motor Vehicle Parts

As shown in table 5.2, U.S. imports of remanufactured goods are estimated to have increased by 21 percent during 2009–11 to nearly $1.5 billion. Nevertheless, U.S. imports of remanufactured motor vehicle parts are estimated to have accounted for only 1– 2 percent of total U.S. imports of motor vehicle parts (new and remanufactured) during the same period ($91.4 billion in 2011).[46] Mexico, the EU, and Canada were the leading suppliers of such imports in 2011 (table 5.6), reflecting their prominent positions as regional motor vehicle parts manufacturers. Leading U.S. firms, such as Cardone and Remy, have remanufacturing facilities in Mexico that supply the U.S. market. In Germany, motor vehicle parts firms Bosch and ZF are active in remanufacturing. Between 2009 and 2011, U.S. imports by foreign remanufacturers invested in the United States are estimated to have increased by 41 percent, from $470 million in 2009 to $664 million in 2011; such imports accounted for 45 percent of total U.S. imports of remanufactured motor vehicle parts in 2011.[47] Most of these imports were likely from the EU and Japan.

Table 5.6. Motor vehicle parts: U.S. imports of remanufactured goods by leading source, 2011

Leading source	2011	% share
	Thousand $	
Mexico	[a]416,420	28.1
EU	[b]314,962	21.3
Canada	[a]37,249	2.5
All other countries or markets	713,308	48.1
Total	1,481,939	100.0

Source: USITC staff calculations of weighted responses to the Commission questionnaire.

Note: U.S. imports from "all other countries or markets" are calculated. Totals may not sum due to rounding.

[a]Low-precision estimate; RSE below 65 percent. [b]Low-precision estimate; RSE below 100 percent.

Trade in Motor Vehicle Parts Cores

Exports of Motor Vehicle Parts Cores

The United States is a net exporter of cores. Between 2009 and 2011, U.S. core exports are estimated to have increased by 32 percent from $148 million in 2009 to $195 million in

2011 (figure 5.1). FTA partner countries are estimated to have accounted for at least 30 percent ($60 million) of U.S. core exports in 2011 (table 5.7). U.S. core exports by foreign remanufacturers invested in the United States are estimated to have accounted for about 63 percent ($123 million) of total U.S. core exports in 2011,[48] with most of these likely destined to the EU.

Imports of Motor Vehicle Parts Cores

Between 2009 and 2011, U.S. core imports are estimated to have nearly tripled to $58 million. FTA partners are estimated to have accounted for 63 percent ($36 million) of total U.S. core imports in 2011 (table 5.8). Detroit Diesel, for example, noted that the movement of remanufactured goods and cores from Australia, an FTA partner, is relatively efficient, and that a significant amount of cores are shipped back by the company to the United States from Australia. [49] U.S. core imports by foreign remanufacturers invested in the United States are considered to be minimal.

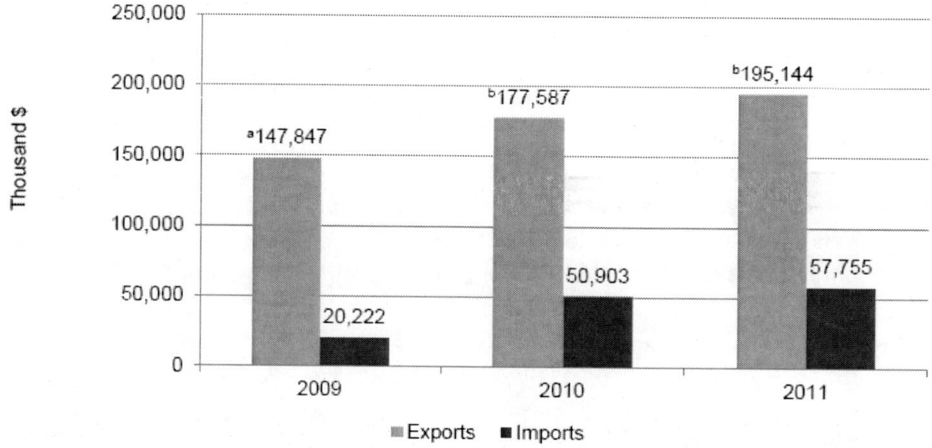

Source: USITC staff calculations of weighted responses to the Commission questionnaire.
[a]Low-precision estimate; RSE below 65 percent. [b]Low-precision estimate; RSE below 80 percent.

Figure 5.1. Motor vehicle parts: U.S. core exports and imports, 2009–11.

Factors Affecting Trade in Remanufactured Motor Vehicle Parts

U.S. exports of remanufactured motor vehicle parts are considered to be small, in part because of barriers that restrict market access. Trade barriers to remanufactured motor vehicle parts are found in numerous countries (table 5.9). Some of these barriers exist due to lack of government policy specific to remanufactured goods.[50]

For example, as noted earlier, few countries have legally defined remanufacturing. Moreover, in many countries, remanufactured goods are often categorized as used products, the importation of which may be restricted or banned.[51]

Bans on imports of used goods may also prevent a remanufacturer from importing the cores required to establish a remanufacturing business in a new market. [52] Other barriers exist as an attempt to protect domestic industries of either new or remanufactured products.[53]

Table 5.7. Motor vehicle parts: U.S. core exports to FTA versus non-FTA partners, 2011

Export market	2011	% share
	Thousand $	
FTA partners	60,217	30.9
Non-FTA partners	[a]7,013	3.6 65.5
Unable to determine[b]	127,914	
Total	[c]195,144	100.0

Source: USITC staff calculations of weighted responses to the Commission questionnaire.

Note: USITC staff was unable to estimate U.S. exports of cores by leading destination because of the low statistical significance of the weighted responses to the Commission questionnaire. Instead, USITC staff estimated U.S. exports of cores to FTA and non-FTA partners as two separate groups, which had higher statistical significance. Totals may not sum due to rounding.

[a]Low-precision estimate; RSE below 65 percent.

[b]Questionnaire respondents were asked to report the value and destination of their core exports. In some instances, respondents reported the value, but not the destination. In these cases, USITC staff was unable to determine the destination of U.S. core exports.

[c]Low-precision estimate; RSE below 80 percent.

Table 5.8. Motor vehicle parts: U.S. core imports from FTA versus non-FTA partners, 2011

Leading source	2011	% share
	Thousand $	
FTA partners	36,160	62.6
Non-FTA partners	[a]21,237	36.8
Unable to determine[b]	358	0.6
Total	57,756	100.0

Source: USITC staff calculations of weighted responses to the Commission questionnaire.

Note: USITC staff was unable to estimate U.S. imports of cores by leading destination because of the low statistical significance of the weighted responses to the Commission questionnaire. Instead, USITC staff estimated U.S. imports of cores to FTA and non-FTA partners as two separate groups, which had higher statistical significance. Totals may not sum due to rounding.

[a]Low-precision estimate; RSE below 65 percent.

[b]Questionnaire respondents were asked to report the value and source of their core imports. In some instances, respondents reported the value, but not the source. In these cases, USITC staff was unable to determine the source of U.S. core imports.

Industry sources also reported that the remanufacturing of motor vehicle parts is generally performed in close proximity to the market of sale, in part because of transportation costs, country-specific regulations, and customer turnaround requirements.[54] Local remanufacturing of components is more cost effective because of the expenses associated with transporting cores.[55] Moreover, because of the costs of shipping cores, trade in these products often depends on the value of the core, with higher-value cores more likely to be traded.[56] Trade in cores may also be restricted by certain countries due to concerns that cores will be resold in those markets and compete against domestic industries that produce new goods. In addition, core exports may be hampered by stringent export documentation requirements that make the process cumbersome for the core owners, who must export their cores to receive a refund of their core deposits or avoid a penalty for not returning usable cores.[57]

Table 5.9. Barriers to U.S. trade in remanufactured motor vehicle parts in selected countries

Country	Trade barrier
Argentina	Bars imports of remanufactured goods, with the exception of remanufactured parts marketed by OEM assemblers to service their own products.
Brazil	Counts cores as used products, and has barriers in place to prevent the importation of used products. Internal barriers to remanufacturing include a vehicle registration system that also requires the engine serial number to be registered. In order to sell remanufactured cores, the producer must convince the person who is the source of the core to go through the process of de-registering his or her old engine.
Chile	Only allows remanufactured goods to be sold on an "outright" basis, without any discount for exchanged cores. U.S. FTA provisions do not address this reported barrier.
China	Restricts remanufactured good imports, and limits domestic remanufacturing. Has allowed some remanufacturing on a "pilot" basis. Allows imports of cores of certain motor vehicle parts into export processing zones (EPZs) for remanufacturing and subsequent export.
India	Does not allow remanufactured products made with cores from other countries to enter its domestic market, but will allow remanufacturing using domestic cores. Allows imports of cores that are intended to be remanufactured domestically and subsequently exported.
Malaysia	Prohibits imports of used automotive parts, including remanufactured goods.
Mexico	Does not include used cores in rules of origin for NAFTA, because they were not included in rules of origin for the original NAFTA agreement.
Russia	Customs regulations make the export of cores difficult.
Turkey	Bars importation of remanufactured products.

Sources: USDOC, ITA, "Compilation of Foreign Motor Vehicle Import Requirements," July 2008, 12; Pierce, written testimony to the USITC, February 9, 2012; U.S. Department of State, U.S. Embassy, Beijing, "China's Remanufacturing Market," February 12, 2012; USITC, hearing transcript, February 28, 2012, 165–66 (testimony of Allen W. Pierce, Cummins Inc.); McKenna, written testimony to the USITC, February 28, 2012.

FDI in the Motor Vehicle Parts Remanufacturing Sector

Inbound FBI in U.S. Remanufacturing Activities

Between 2009 and 2011, estimated FDI flows in U.S. remanufacturing activities swung widely, likely reflecting one-time acquisitions or large investments in remanufacturing activities by one or two firms (table 5.10).

Foreign manufacturers that produce new motor vehicle parts in the United States or import these products into the United States are also sources of FDI in U.S. remanufacturing.[58]

Table 5.10. Motor vehicle parts: FDI and associated employment in U.S. remanufacturing, 2009–11

Item	2009	2010	2011	% change, 2009–11
	Thousand $			
Inbound FDI	[a]201,904	40,084	[a]15,237	-92.5
Outbound FDI	11,567	9,244	13,616	17.7
	Full-time workers			
U.S. remanufacturing employment associated with inbound FDI	3,541	3,748	3,576	1.0

Source: USITC staff calculations of weighted responses to the Commission questionnaire.
[a]Low-precision estimate; RSE below 65 percent.

For example, Aisin, a Japanese OEM transmission supplier, remanufactures transmissions for the U.S. market through a subsidiary in Michigan called AWTEC.[59]

For some other companies, the production of new motor vehicle parts in the United States may not make economic sense, whereas the case for remanufacturing motor vehicle parts may be stronger due to the savings in transportation costs.

U.S. remanufacturing employment associated with inbound FDI was estimated to be about 3,600 during the period. Employment associated with FDI in remanufacturing encompasses both companies that remanufacture in the United States and those that import remanufactured motor vehicle parts into the United States.

Outbound FBI in Foreign Remanufacturing Activities

Between 2009 and 2011, U.S. investment in foreign remanufacturing activities is estimated to have increased modestly to almost $14 million in 2011 (table 5.10). The motor vehicle remanufacturing supply chain tends to be regional rather than global, with the same individual or company acting as both a consumer and a supplier of cores. U.S. companies that invest in foreign remanufacturing facilities typically do so to pursue lower costs or access to a new market.

Mexico, for example, is often a destination for U.S. remanufacturing investment, generally as a means to reduce remanufacturing costs through lower labor costs and lower transportation costs due to the close proximity to the U.S. market.

Because transportation costs are high, production and sale of remanufactured goods reportedly tend to be most cost effective if performed in the same region.[60] Cummins, Detroit Diesel, and Visteon are among the OEMs that have remanufacturing facilities in Mexico.[61] Independent remanufacturers Cardone and Remy International also remanufacture motor vehicle parts in Mexico for export to the U.S. market.[62]

FDI is also reportedly used to gain access to a market where imports of remanufactured goods are prohibited.[63] In that case, an OES may couple a remanufacturing plant with a manufacturing plant.

For example, Cummins India announced a $300 million investment in a three-factory project, including a manufacturing and remanufacturing plant for diesel engines, in January 2011.[64]

REFERENCES

Automotive Parts Remanufacturing Association (APRA). "What Is Remanufacturing?" n.d. http://www.apra.org/About/Reman.asp (accessed February 16, 2012).

Clottey, Toyin, and W.C. Benton. "Core Acquisitions Planning in the Automotive Parts Remanufacturing Industry." *The Reman Institute,* 2010. http://www.reman.org /Publications_main.htm.

Cummins Inc. "Cummins Megasite Opens in India." News release, January 2011.

———. "Remanufacturing is the New 'New.'" *Sustainability Report 2010–2011*, 2011, 14–18.

Gager, William C. Automotive Parts Remanufacturing Association (APRA). Written testimony submitted to the U.S. International Trade Commission in connection with inv. no. 332-525, *Remanufactured Goods: An Overview of the U.S. and Global Industries, Markets, and Trade,* February 28, 2012.

Goebel, Dave. "The Changing Auto Industry—The Impact of Declining New Vehicle Sales on the Aftermarket Business." *Polk Market Study,* March 2009, 7.

IBIS. *Used Car Parts Wholesaling in the US.* IBISWorld Industry Report 42114, March 2011.

Kekre, Sunder, Uday S. Rao, Jayashankar M. Swaminathan, and Jun Zhang. "Reconfiguring a Remanufacturing Line at Visteon, Mexico." *Interfaces,* November–December 2003, 30–43.

Lund, Robert T., and Hauser, William M. *Remanufacturing: An American Perspective.* Boston University, n.d. http://www.reman.org/Papers /ICRM%20APRA%20version.pdf (accessed August 23, 2012).

McKenna, Robert E. Motor & Equipment Manufacturers Association (MEMA). Written testimony submitted to the U.S. International Trade Commission in connection with inv. no. 332-525, *Remanufactured Goods: An Overview of the U.S. and Global Industries, Markets and Trade,* February 28, 2012.

Meritor, Inc. "Meritor's Environmentally Friendly Remanufacturing Efforts Promote Fleet Efficiency," June 28, 2012. http://www.mera.org /Communications/Remanufacturing-News/June-28-2012/Meritors-Remanufacturing-Promotes-Fleet-Efficiency.html.

Pierce, Allen W. Cummins Inc. Written testimony submitted to the U.S. International Trade Commission in connection with inv. no. 332-525, *Remanufactured Goods: An Overview of the U.S. and Global Industries, Markets, and Trade,* February 9, 2012.

Schau, Erwin M., Marzia Traverso, Annekatrin Lehmann, and Matthias Finkbeiner. "Life Cycle Costing in Sustainability Assessment: A Case Study of Remanufactured Alternators." *Sustainability,* November 18, 2011.

Schinzing, Russ. "Cores—Cores—Cores." Electronic Remanufacturing Company, July 2007. http://e-reman.com/blog/cores-cores-cores.

Sevart, Matthew. "Certified Transmission Thrives with New Distributors and Adds 100 Employees." *APRA Global Connection,* March 2012.

Shephardson, David. "Average Age of Cars, Trucks Hits 10.8 years, Sets Record." *Detroit News,* January 18, 2012. http://www.detroitnews.com /article/20120118 /AUTO01 /201180329.

U.S. Department of Commerce (USDOC). International Trade Administration (ITA). "Compilation of Foreign Motor Vehicle Import Requirements," July 2008. http://trade.gov/wcm/groups/internet/@trade/@mas/@man /@aai/documents/web_content/auto_re ports tradebarriers.pdf.

———. "On the Road: U.S. Automotive Parts Industry Annual Assessment," 2010.

U.S. Department of State. U.S. Embassy, Beijing, "China's Remanufacturing Market (BEIJING 000460)," February 12, 2012.

U.S. International Trade Commission (USITC). Hearing transcript in connection with inv. no. 332-525, Remanufactured Goods: An Overview of the U.S. and Global Industries, Markets, and Trade, February 28, 2012.

INFORMATION TECHNOLOGY (IT) PRODUCTS

Description of IT Product Remanufacturing Activities

Broadly defined, the sector encompassing remanufactured IT and imaging equipment (collectively, "remanufactured IT products") includes firms that remanufacture computer hardware, telecommunications (telecom) network equipment, and imaging products.[1] Examples of computer hardware include personal computers (PCs), servers, motherboards, and hard disk drives. Telecom network equipment typically includes circuit packs, modems, routers, and base stations. Imaging products include home and office multifunction printers, digital printing presses, and parts, including toner cartridges.

The remanufacturing process for IT products varies, given the diversity of products in this sector. In general, remanufacturers acquire used IT products and determine which items are worth remanufacturing, as opposed to applying other, less extensive processes, such as refurbishing or recycling (box 6.1). Remanufactured equipment is repaired and restored to original working condition using processes as similar as possible to those used to produce an equivalent new product. Remanufactured IT products are then sold to customers, either directly or through a distribution system.

Box 6.1. IT products: Remanufacturing versus refurbishing

Both remanufacturing and refurbishing occur in the IT sector, although the differences between the two processes can be a matter of nuance, and some firms in the IT sector use the terms interchangeably.

As the definition provided in the USTR's request letter to the Commission states, remanufactured goods are "entirely or partially comprised of parts that have been obtained from the disassembly of used goods."

The refurbishing process for IT products, on the other hand, tends not to involve disassembly: certain components may be cleaned or replaced with new parts, but the original item remains largely intact. Both remanufactured and refurbished products usually undergo thorough testing, and both tend to come with warranties, although the warranty on a refurbished good may be more limited than the warranty on a remanufactured product.

An example of the difference between the two comes from one firm in the imaging industry that both remanufactures and refurbishes its printing machines.

The firm's refurbished machines have undergone light maintenance and upgrades, but have not been disassembled. Instead, refurbished machines are sold as "used" for substantially less than a new machine, and come with a one-year warranty. Remanufactured machines, on the other hand, are disassembled, and all high-wear components are replaced.

The performance and price are the same as for a new machine, and a warranty similar to that of a new machine is offered.[a]

The decision about whether to remanufacture an IT product, refurbish it, or recycle its parts relates to the product life cycle. Products that are remanufactured tend to be products with longer life cycles, in some cases 20 years or more.

Examples include digital printing presses and circuit packs for telecom networks. Refurbishing is more common for consumer-oriented IT products, such as PCs and home printers, which have shorter life cycles and declining selling prices for the new product: investment in remanufacturing is less likely to be economically rational in these cases. Products that are not reusable in their current state, whether due to low demand or poor condition, are recycled, often with some high-value parts or materials being salvaged and sold.

[a]Industry representative, telephone interview by USITC staff, November 1, 2011.

Size of IT Remanufacturing in the United States

Industry Structure

The IT remanufacturing sector in the United States is relatively fragmented and likely consists of thousands of firms, although most are small: there are estimated to be fewer than 500 firms that employ more than 20 workers.[2] According to the International Imaging Technology Council (IITC), a trade association representing third-party printer cartridge remanufacturers, there are approximately 2,000 firms just in the printer cartridge remanufacturing sector in the United States.[3] Both OEMs and third-party firms remanufacture IT equipment, although OEMs are believed to account for a larger share of this activity. Still, remanufacturing reportedly accounts for a very small fraction of large OEM operations and, by extension, of employment at these firms. According to one large firm in the telecom equipment industry, even though it was among the first OEMs in the industry to remanufacture IT products, remanufacturing activities still only account for about 1 percent of company revenue.[4] Indeed, remanufactured IT products are estimated to account for no more than 0.5 percent of total sales of IT products in the United States.[5]

Third-party firms may perform contract remanufacturing for OEMs, or they may purchase IT equipment cores and remanufacture and sell IT products independently, thus competing directly with OEMs. Third-party toner cartridge remanufacturers, for example, are predominantly SMEs, and often compete directly with OEM cartridge remanufacturers. There are more small firms involved in remanufacturing IT products than there are in producing new ones. For example, the PC and toner cartridge remanufacturing segments are dispersed among a large number of small firms, while the production of new PCs and cartridges is very concentrated. It is more common for firms to sell both remanufactured goods and new goods than it is for firms to produce both of them. It is estimated that less than 40 percent of U.S. remanufacturers of IT equipment produce both remanufactured goods and new goods, but that almost 90 percent of firms sell both.[6]

Remanufactured IT products are often produced in close proximity to the markets they serve. This helps to reduce the transportation costs of sourcing cores and shipping finished products and also takes advantage of speed to market. As a result, exports also tend to be regional. In comparison, the supply chains for new IT product manufacturing are more global in scope. Moreover, even as the production of new IT products is concentrated in Asia,[7] exports of new IT products tend to be more global. However, some larger firms may attempt to integrate both IT manufacturing and remanufacturing operations in order to leverage existing infrastructure. For instance, some remanufacturing operations may be performed in the same facility as new product manufacturing, or close by.[8]

Production, Investment, and Employment

Between 2009 and 2011, annual production of remanufactured IT products is estimated to have remained essentially unchanged at $2.6–$2.7 billion (table 6.1). Remanufactured IT products are estimated to have accounted for less than 1 percent of total sales by U.S. firms within the IT products sector during the period ($727 billion in 2011).

Table 6.1. IT products: Remanufacturing production, investment, and employment, 2009–11

Item	2009	2010	2011	% change, 2009–11
	Thousand $			
Production	2,709,170	2,592,831	2,681,603	-1.0
Investment	14,892	15,950	17,503	17.5
	Full-time workers			
Employment	11,493	13,025	15,442	34.4

Source: USITC staff calculations of weighted responses to the Commission questionnaire.

U.S. investment in IT remanufacturing is estimated to have increased by 18 percent from $14.9 million to $17.5 million during the period. The reported low levels of U.S. investment in IT remanufacturing, such as in R&D or property, plant, and equipment, likely reflect the narrowing price difference between new and remanufactured IT products. The small size of the price gap reduces consumers' incentive to buy remanufactured IT products, depressing demand and undermining incentives to invest in additional production capacity. In fact, remanufacturing production capacity is estimated to have remained unchanged for 44 percent of IT remanufacturers, although an estimated 33 percent of IT remanufacturers expanded their production capacity during the period. The divergence may reflect a difference in product mix and demand. About 5 percent of IT remanufacturers are estimated to have reduced their capacity at existing facilities, while very few are estimated to have either acquired new facilities or divested assets.[9]

The Commission estimates that U.S. employment in the IT remanufacturing sector increased by 34 percent from about 11,500 in 2009 to 15,500 in 2011. These estimates may, however, be low: one industry representative estimated U.S. employment in the U.S. printer cartridge remanufacturing industry at 33,650.[10]

U.S. Market for Remanufactured IT Products

Market Size

Between 2009 and 2011, the U.S. market for remanufactured IT products is estimated to have increased by 17 percent from $4.4 billion in 2009 to $5.2 billion in 2011 (table 6.2). Remanufacturers export only about 10 percent of production, reflecting their focus on the domestic market. In 2011, the value of imports of remanufactured IT products exceeded the value of domestic production.

Table 6.2. IT products: U.S. market (apparent consumption) for remanufactured goods, 2009–11

Item	2009	2010	2011	% change, 2009–11
		Thousand $		
Production	2,709,170	2,592,831	2,681,603	-1.0
Imports	1,931,994	2,281,198	2,756,475	42.7
Exports	219,961	250,197	260,032	18.2
U.S. apparent consumption	4,421,204	4,623,832	5,178,046	17.1

Source: USITC staff calculations of weighted responses to the Commission questionnaire. Note: Totals may not sum due to rounding.

U.S. Buyers and Demand Factors

Buyers of remanufactured IT products are diverse. Retail stores buy and sell remanufactured computer equipment, printers, and toner cartridges. Some larger outlets maintain sections devoted exclusively to remanufactured notebook computers, allowing the customer to comparison-shop between remanufactured and new computers. Telecom network providers principally buy (or lease) remanufactured network equipment. Other buyers of remanufactured network equipment include government, educational institutions, banks, and hospitals.

Demand for remanufactured IT products is primarily based on customer perception and price. Customers must be convinced that the remanufactured product will perform adequately, and the likelihood of this happening varies depending on the product in question as well as the characteristics of the buyers. Price also drives demand, as buyer interest in remanufactured IT equipment depends heavily on the price of competing new IT products. As noted above, some remanufactured IT equipment, such as PCs, face competition from new goods that may be technologically superior and consistently declining in price owing to innovation and shorter product life cycles.

The share of remanufactured goods in the broader U.S. market for IT products varies. In some cases, such as with digital printing presses, many buyers recognize remanufactured products as being fully interchangeable with new products, so remanufactured goods account for a larger share of overall sales.[11] For telecommunications equipment, network providers may stipulate in contracts that they can substitute remanufactured products for new ones at their discretion.[12] For printer cartridges, remanufactured sales range from 10 to 50 percent of the overall replacement cartridge market, depending on the printer model. OEMs account for the majority of these sales, but third-party firms are important players as well.[13]

As noted earlier, remanufacturing is somewhat countercyclical, and during the years of slow growth following the 2008–09 economic recession, consumers may have switched to buying remanufactured IT products instead of new ones to save money. However, this may be less true for this sector than for other remanufactured goods. The option of buying low-cost new IT products may prompt consumers to decide against remanufactured ones. It is estimated that consumer preferences for new IT products, availability of low-cost new IT products, and declining demand for remanufactured IT products affect the competitive position of over 20 percent of the IT remanufacturers in the United States.[14]

U.S. Sellers and Supply Factors

Sellers of remanufactured IT products include OEMs and independent remanufacturers, distributors, and retailers. The supply of remanufactured IT equipment is driven by the cost of remanufacturing. This cost is set by several factors, including the availability and price of IT cores, the cost of sorting and processing used goods to determine their suitability for remanufacturing, labor costs, and transportation costs. These factors make remanufacturing a considerable undertaking for firms and determine what products will be remanufactured and in what quantities.

IT equipment remanufacturers maintain relationships with a variety of core suppliers, which include brokers and distributors, recyclers, and existing customers (via returns, rebates, or exchange programs). Distributors and brokers are commonly used by third-party printer cartridge remanufacturers. Electronics recyclers often determine which of the used IT products that they receive can be resold (versus those that should be recycled), and their customers can include refurbishers or remanufacturers. This service is highly valued because the quality of used IT products can vary significantly and is difficult to assess, directly impacting the potential market value of the remanufactured product.[15]

Sorting and processing costs affect the overall cost of remanufacturing. If the incoming supply of used products has not been presorted, firms must sort them into scrap, products to sell as used or refurbished goods, and products to remanufacture. For most IT products, all three of these streams exist for used products. One OEM estimates that it scraps about 50 percent of core returns, while another OEM reported more variable scrap rates of core returns depending on the product and end user.[16]

Remanufacturing is often more labor-intensive than new product manufacturing, and as a result, labor costs are a major consideration for most firms and drive decisions about whether and where to remanufacture a product. While only an estimated 10 percent of IT remanufacturers rank labor costs as an extremely important competitive factor,[17] the importance of Mexico as a remanufacturing location for IT products (described below) is one indicator of the attention firms pay to balancing labor costs with other competitive factors. Finally, the costs of shipping inbound cores to be processed, and of shipping outbound remanufactured goods for distribution, affect the cost of remanufacturing. The desire to minimize shipping costs is a major reason why, as noted previously, remanufacturing is more often located close to the market served as opposed to having one site for worldwide remanufacturing activities.

Sales patterns for remanufactured IT products, particularly for printer toner cartridges, may be affected by the competitive relationship between OEM remanufacturers and third-party remanufacturers. According to some industry participants, OEMs engage in various tactics to compete with third-party firms, including public-relations campaigns to discredit the quality of third-party remanufactured cartridges and efforts to limit the supply of cartridge cores. These efforts reportedly include making cartridges more technologically complex and thus more difficult for third-party firms to remanufacture, and promoting discount programs directly with customers to exchange new cartridges at a discount for used ones.[18] OEMs have also implied that using non-OEM cartridges will void a printer's warranty, even though these types of claims have been ruled to be in violation of the Magnuson-Moss Warranty Act.[19]

Estimates of U.S. Trade in Remanufactured IT Products and Cores

Trade in Remanufactured IT Products

Exports of Remanufactured IT Products

The United States is a net importer of remanufactured IT products, as it exports only a small portion of the remanufactured IT equipment it produces (see table 6.2). Moreover, while U.S. exports of remanufactured IT products increased by approximately 18 percent over the study period from $220 million in 2009 to $260 million in 2011 (table 6.3), they are estimated to have accounted for less than 1 percent of total U.S. exports of all IT products (new and remanufactured) during the period ($69.3 billion in 2011).[20] The EU and Canada are estimated to be the destination for about 45 percent of U.S. exports of remanufactured IT products.

Between 2009 and 2011, U.S. exports of remanufactured IT products to FTA partners are estimated to have increased by 12 percent, from $89 million in 2009 to $99 million in 2011 (table 6.4).

Table 6.3. IT products: U.S. exports of remanufactured goods by leading destination, 2009–11

Country or market	2009	2010	2011	% change, 2009–11
		Thousand $		
EU	[a]31,304	[b]64,287	[a]63,813	103.8
Canada	54,240	52,979	54,880	1.2
Hong Kong	[c]41,857	[c]33,488	[c]25,993	–37.9
All other countries or markets	92,560	99,443	115,346	24.6
Total	219,961	250,197	260,032	18.2

Source: USITC staff calculations of weighted responses to the Commission questionnaire. Note: Totals may not sum due to rounding.

[a]Low-precision estimate; RSE below 65 percent.

[b]Low-precision estimate; RSE below 80 percent.

[c]Low-precision estimate; RSE below 100 percent.

Table 6.4. IT products: U.S. exports of remanufactured goods to FTA partners, 2009–11

FTA partner	2009	2010	2011	% change, 2009–11
		Thousand $		
Canada	54,240	52,979	54,880	1.2
Mexico	19,387	27,745	20,519	5.8
Singapore	7,248	6,565	9,878	36.3
All other FTA partners	7,662	10,243	13,639	78.0
Total FTA partner exports	88,537	97,532	98,916	11.7

Source: USITC staff calculations of weighted responses to the Commission questionnaire. Note: Totals may not sum due to rounding.

FTA partners are estimated to have accounted for 38–40 percent of total U.S. exports of remanufactured IT products in this period, with the majority of such exports destined for Canada and Mexico. U.S. exports of remanufactured IT equipment by foreign remanufacturers invested in the United States, primarily to Canada and Japan, are estimated to have accounted for about 7 percent ($19 million) of total U.S. exports of remanufacturing IT equipment during 2011 (table 6.5).

Imports of Remanufactured IT Products

U.S. imports of remanufactured IT products totaled nearly $2.8 billion in 2011, but are estimated to have accounted for only 1–2 percent of total U.S. imports of all IT products (new and remanufactured) during the period ($186.2 billion in 2011).[21] While Mexico is also an important assembly location for certain segments of the new product manufacturing industry in the IT sector, its role is even more prominent in remanufacturing. In fact, Mexico by far is the leading supplier of remanufactured IT products to the U.S. market (table 6.6). Mexico's proximity to the U.S. market helps firms handle the complex logistics of the remanufacturing process and also helps in keeping transportation costs low.

Table 6.5. IT products: U.S. exports of remanufactured goods by foreign remanufacturers invested in the United States by leading destination, 2009–11

Country	2009	2010	2011	% change, 2009–11
		Thousand $		
Canada	[a]6,289	[a]6,655	[a]6,792	8.0
Japan	[b]5,001	[b]5,880	[b]4,594	–8.1
Mexico	([c])	([c])	[b]2,373	n/a
All other countries	[a]5,286	[a]6,533	[a]4,933	–6.7
Total	16,575	19,067	18,692	12.8

Source: USITC staff calculations of weighted responses to the Commission questionnaire. Note: Totals may not sum due to rounding. N/a = not applicable.

[a]Low-precision estimate; RSE below 65 percent.

[b]Low-precision estimate; RSE below 80 percent.

[c]None reported.

Table 6.6. IT products: U.S. imports of remanufactured goods by leading source, 2011

Country	2011	% share
	Thousand $	
Mexico	838,960	30.4
All other countries	1,917,514	69.6
Total	2,756,475	100.0

Source: USITC staff calculations of weighted responses to the Commission questionnaire.

Note: Figures for U.S. imports from "all other countries" are calculated. Totals may not sum due to rounding.

Mexico offers a labor cost advantage over the United States while still offering relatively low transportation costs, especially given that most IT remanufacturing is done in free trade zone (FTZ) facilities that tend to be located near the U.S. border.[22] Because remanufactured products compete with new goods that are, in many cases, declining in price and/or that have improved technological features, speed to market is an important competitive factor. The faster these products can be refurbished or remanufactured and put back up for sale, the more likely it is that remanufacturing will be profitable.[23] Data for U.S. imports of remanufactured IT products by foreign remanufacturers invested in the United States are not available.[24]

U.S. Trade in IT Cores

Exports of IT Cores

The United States is a large net exporter of cores (figure 6.1). Exports to FTA partners are estimated to have accounted for practically all of U.S. core exports in 2011 (table 6.7). The total value of core exports ($507 million) exceeded the value of exports of finished remanufactured IT products ($260 million) in 2011. This is likely explained by at least two factors. First, the United States is a very large market for IT products, and thus generates a large number of cores. Second, to keep transportation costs low, some IT remanufacturers tend to ship cores to nearby countries to conduct remanufacturing and then bring the remanufactured product back to the United States—a major reason for Mexico's prominence as a source of finished remanufactured goods. Exports of cores by foreign remanufacturers invested in the United States are estimated to have accounted for 41 percent ($207 million) of total U.S. IT core exports in 2011.

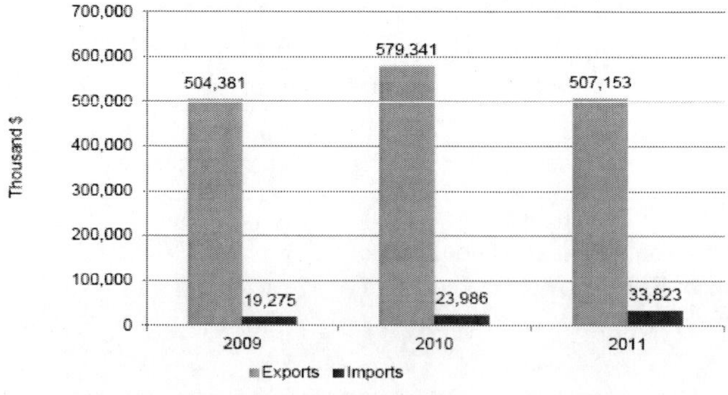

Figure 6.1. IT products: U.S. core exports and imports, 2009–11.

Imports of IT Cores

U.S. imports of IT cores totaled almost $34 million in 2011, and about one-half of these imports came from FTA partner countries (table 6.8). In contrast to the situation for exports, where exports of cores exceeded exports of finished remanufactured products, imports of cores were much smaller than imports of finished remanufactured goods. Imports of cores by foreign remanufacturers invested in the United States were quite small in value, at around $3 million, and accounted for a little less than 10 percent of total U.S. imports of IT cores.[25]

Table 6.7. IT products: U.S. core exports to FTA versus non-FTA partners, 2011

	2011	% share
Export market	Thousand $	
FTA partners	499,584	98.5
Non-FTA partners	[a]3,811	0.8
Unable to determine[b]	3,758	0.7
Total	507,153	100.0

Source: USITC staff calculations of weighted responses to the Commission questionnaire.

Note: USITC staff was unable to estimate U.S. exports of cores by leading destination because of the low statistical significance of the weighted responses to the Commission questionnaire. Instead, USITC staff estimated U.S. exports of cores to FTA and non-FTA partners as two separate groups, which had higher statistical significance. Totals may not sum due to rounding.

[a]Low-precision estimate; RSE below 65 percent.

[b]Questionnaire respondents were asked to report the value and destination of their core exports. In some instances, respondents reported the value, but not the destination. In these cases, USITC staff was unable to determine the destination of U.S. core exports.

Table 6.8. It products: U.S. core imports from FTA versus non-FTA countries, 2011

	Thousand $	
Leading source	2011	% share
FTA countries	16,610	49.1
Non-FTA countries	9,629	28.5
Unable to determine[a]	7,584	22.4
Total	33,823	100.0

Source: USITC staff calculations of weighted responses to the Commission questionnaire.

Note: USITC staff was unable to estimate U.S. imports of cores by leading source because of the low statistical significance of the weighted responses to the Commission questionnaire. Instead, USITC staff estimated U.S. imports of cores from FTA and non-FTA partners as two separate groups, which had higher statistical significance. Totals may not sum due to rounding.

[a]Questionnaire respondents were asked to report the value and source of their core imports. In some instances, respondents reported the value, but not the source. In these cases, USITC staff was unable to determine the source of U.S. core imports.

Factors Affecting Trade in Remanufactured IT Products

As previously noted, U.S. remanufacturing of IT products tends to be focused on production for the domestic market. The small share of production that is exported tends to be to nearby countries. Producers' ability to export is also limited by regulations. Some countries limit or prohibit imports of remanufactured IT products entirely. For example, China heavily restricts imports of used electronics, and an import permit is required.[26]

In Vietnam, remanufactured IT hardware may be imported only under certain conditions, such as when an item is intended for re-export or a repaired item is being returned to its owner.[27] Government regulations for remanufactured IT products can also reinforce the regional production model of remanufacturing.

For instance, one large IT remanufacturer reported separating remanufacturing product streams for the U.S. versus EU markets to ensure compliance with the EU's Restriction of the Use of Certain Hazardous Substances (RoHS) directive for the EU market.[28]

RoHS is of particular importance because IT products are likely to contain substances that are restricted under the directive. For example, lead, a restricted substance, is commonly used in soldering for electronics products, including on printed circuit boards. An estimated 18 percent of IT remanufacturers consider environmental regulations as an extremely important factor affecting their ability to compete in global markets.[29] Imports of remanufactured toner cartridges from Asia are reportedly increasing. According to the IITC, these imports include both counterfeit products and legitimate remanufactured cartridges. Such imports, combined with the OEM measures to control the supply of cores, have reportedly helped reduce U.S. cartridge remanufacturers' share of world production from one-half to one-third.[30]

FDI in the IT Products Remanufacturing Sector

Figures for FDI in IT remanufacturing activities in the United States and for U.S. investment in foreign remanufacturing activities are not readily available. [31] U.S. employment by foreign IT products remanufacturers invested in the United States is estimated to account for about 9 percent (1,800 workers) of total U.S. remanufacturing employment in the IT remanufacturing sector in 2011, down from 18 percent (2,500 workers) in 2009. U.S. investment in foreign remanufacturing activities is likely directed to Mexico, given the country's existing remanufacturing facilities and proximity to the U.S. market.

References

Bernstein Research. "Remanufactured Cartridges." In Lexmark: Amid the Controversy, an Attractive Business Model and Valuation, July 2005.

Government of Vietnam. Ministry of Information and Communications. Circular Promulgating a List of Used Information Technology Appliances Banned from Import. No. 43/2009-TT-BTTTT, December 30, 2009.

Judge, Tricia. Written testimony submitted to the U.S. International Trade Commission in connection with inv. no. 332-525, Remanufactured Goods: An Overview of the U.S. and Global Industries, Markets, and Trade, February 9, 2012.

Marcotte, Suzanne, Marie-Eve Hallé, and Benoit Montreuil. "Computer Hardware Reverse Logistics: A Field Study of Canadian Facilities." Interuniversity Research Centre on

Enterprise Networks, Logistics, and Transportation, August 2008. https://www.cirrelt.ca/DocumentsTravail /CIRRELT-2008-41.pdf.
UK Department for Business Innovation and Skills. "What Is RoHS?" http://www.bis.gov.uk/nmo/enforcement/rohs-home (accessed April 30, 2012).
U.S. Federal Trade Commission (FTC). Bureau of Consumer Protection. "Understanding the Magnuson-Moss Warranty Act." http://business.ftc. gov/documents/bus01-business personsguide-federal-warranty-law#Magnuson-Moss (accessed April 30, 2012).

MEDICAL DEVICES

Description of Medical Device Remanufacturing

The remanufactured medical device sector forms part of the broader industry for previously owned (or pre-owned) medical devices and equipment. Pre-owned medical devices commonly fall into three categories—refurbished, remanufactured, and "as is"—based on the extent to which the original device's function is altered and the type of device.

Refurbished medical devices are primarily medical imaging equipment that is restored to OEM or equivalent specifications without changing the equipment's performance, safety specifications, or intended use. Refurbished medical devices that undergo any upgrades are consistent with the original product specifications and service procedures. [1] Remanufactured medical devices, as defined by the U.S. Food and Drug Administration (FDA), undergo restoration processes that significantly change the device's original performance or safety specifications, or intended use.[2] Reprocessed medical devices, also according to the FDA, are remanufactured single-use devices (SUDs), which are described below. [3] Medical devices that are either refurbished or remanufactured/ reprocessed (as defined by the FDA) fall within the scope of the definition of remanufactured goods given to the Commission by the USTR (figure 7.1).[4] "As is" medical devices are generally sold in the same condition in which they were acquired,[5] and meet the definition of cores. For the purposes of this report, the term "remanufactured medical devices" refers to refurbished medical imaging equipment, as well as to both remanufactured medical devices and reprocessed SUDs under the FDA definition.[6]

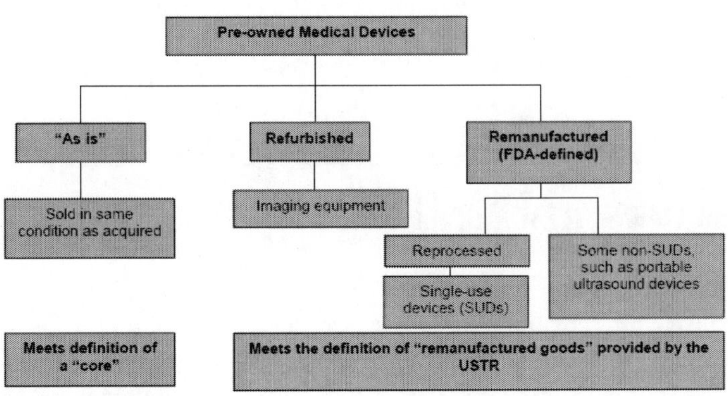

Source: Compiled by USITC staff.

Figure 7.1. Pre-owned medical devices commonly fall into three categories.

Remanufactured medical devices primarily consist of refurbished medical imaging equipment and reprocessed SUDs. Examples of imaging equipment include magnetic resonance imaging (MRI), computed tomography (CT), X-ray, interventional cardiology, and various ultrasound devices.[7] Medical imaging devices are capital intensive both to buy and to use. Due to their advanced technology, they are expensive, costing upwards of $1 million;[8] they also pose a relatively high risk to patients. Conversely, SUDs cost less and represent less risk to patients; they include blood pressure cuffs, diagnostic electrophysiology catheters, forceps, and babcocks (a type of forceps). These devices are frequently used for a variety of procedures, including cardiovascular, arthroscopic/orthopedic, general surgery, gastroenterology, and laparoscopic surgery procedures.[9] The medical device remanufacturing process generally includes selecting the used equipment to be remanufactured, disassembling and inspecting it, restoring it to original working condition or better, reinstalling it, and providing service support (box 7.1).[10]

Box 7.1. The medical device remanufacturing process

Medical imaging equipment. The remanufacturing of this equipment commonly follows five steps.[a] First, the equipment is inspected and tested to determine whether the device is in good enough condition to be remanufactured to its original standards. The criteria for evaluation tend to include the age of the device, service history, past performance, and the extent to which software and hardware need to be upgraded. Next, the device is deinstalled, packaged, and transported to a facility where the device is remanufactured. At this stage, imaging devices are sanitized; inspected, with worn parts replaced; and upgraded with the most appropriate software and hardware. Next, the imaging equipment is transported to the customer, reinstalled, and tested. Finally, in addition to providing warranties, remanufacturers may provide financing solutions and service contracts, as well as spare parts support for the remainder of the product's lifecycle.

SUDs. Once opened or unsterilized in any way, SUDs are deposited into designated containers. The containers are then collected by the reprocessing company so that the SUDs can be cleaned, inspected, and ultimately repaired if needed. Once this process is completed, the devices are repackaged and returned to the end user.[b] SUDs are generally remanufactured up to five times,[c] and when they can no longer be remanufactured, 95 percent of SUDs are reportedly recycled instead of being disposed of, as many of the constituent parts include valuable commodities such as stainless steel, gold, and titanium.[d]

[a]COCIR, Good Refurbishment Practice for Medical Imaging Equipment, October 2009, 10–16.
[b]USITC, hearing transcript, February 28, 2012, 206 (testimony of Dan Vukelich, AMDR).
[c]SterilMed website, "FAQs," n.d. (accessed May 16, 2012).
[d]USITC, hearing transcript, February 28, 2012, 207 (testimony of Dan Vukelich, AMDR).

Size of Medical Device Remanufacturing in the United States

Industry Structure

The U.S. industry for remanufactured medical devices is composed of a handful of large OEMs and hundreds of third-party providers, the majority of whom are SMEs.[11] The emergence of OEMs in this industry is a relatively recent development and represents a considerable shift in the OEMs' strategy, from relying primarily on developing and commercializing new devices towards one that now includes the remanufacturing and remarketing of pre-owned medical devices. [12] Before 2000, third-party providers dominated the industry for remanufactured medical equipment, with mixed results. Providers ranged in quality from highly reputable firms to little-known outfits that did not always adhere to quality standards.[13] Further, remanufacturing by third-party providers commonly voided the OEM's warranty and providers did not furnish additional servicing for the equipment, meaning that end users would have little recourse in the event of a malfunctioning device.[14] Frequently, disgruntled end users would direct their complaints to the OEM.[15]

In an effort to preserve their reputations and capitalize on the growing market for remanufactured equipment, OEMs increasingly began to remanufacture their own medical imaging devices, and entered the SUD remanufacturing segment of the industry as well. For instance, within the past two years, Johnson & Johnson and Stryker Corp., two of the world's largest medical device firms, acquired the most prominent SUD remanufacturing firms in the United States. With the acquisitions, these two OEMs now account for 95 percent of domestic SUD remanufacturing.[16] The situation is somewhat different with respect to medical imaging equipment, as OEMs have authorized some third-party providers to remanufacture their devices.[17] This collaborative arrangement lends credibility to the third-party provider, who is only authorized to remanufacture a device with OEM parts.

However, the influence of OEMs can also limit third parties from entering the market. OEMs may have a distinct advantage in this industry, given their familiarity with their technologies coupled with their infrastructure to service, finance, and remanufacture their own products. Further, OEMs of medical imaging equipment have increasingly resisted sharing their software and technologies with other, non-approved third-party remanufacturers, making it difficult for these firms to service OEM-issued devices.[18] Similarly, OEMs of SUDs have the authority to determine whether their devices are literally "single-use" or in fact reusable, and some industry representatives suggest that the decision to label a device "single-use" is intended to discourage third-party reprocessing.[19]

Production, Investment, and Employment

U.S. production of remanufactured medical devices is estimated to have increased by about 12 percent, from $1.3 billion in 2009 to $1.5 billion in 2011 (table 7.1).[20] During the same period, remanufactured medical devices are estimated to have accounted for less than 1 percent of total sales by U.S. firms within the medical devices sector ($312 billion in 2011). [21] The modest increase in production coincides with an increase in remanufacturing production capacity. It is estimated that nearly half (49 percent) of medical device remanufacturers expanded their production capacity during this time.[22]

Table 7.1. Medical devices: Remanufacturing production, investment, and employment, 2009–11

Item	2009	2010	2011	% change, 2009–11
	Thousand $			
Production	1,307,588	1,367,739	1,463,313	11.9
Investment	21,079	25,184	31,260	48.3
	Full-time workers			
Employment	3,781	3,910	4,117	8.9

Source: USITC staff calculations of weighted responses to the Commission questionnaire.

Domestic investment in the remanufactured medical devices sector, such as capital expenditures in plant construction or R&D, is estimated to have increased by almost 50 percent, from $21 million in 2009 to $31 million in 2011. Employment in the U.S. remanufactured medical devices sector is estimated to have increased slightly, from about 3,800 in 2009 to 4,100 in 2011. In comparison, the broader U.S. medical device industry reportedly employs more than 350,000 workers in the United States.[23]

Factors Affecting Sector Trends in Sales and Investment

Cost-cutting measures and the general shift towards value-driven healthcare have been the principal factors affecting recent trends in sales of remanufactured medical devices.[24] In particular, declining insurance reimbursements for imaging procedures have become more prevalent.[25] End users are commonly reimbursed through third-party insurers— either through private companies or through the government via the Centers for Medicare and Medicaid Services.[26] Given the high cost of these devices, hospitals and imaging centers alike must increasingly demonstrate long-term benefits from their use. However, many insurers associate high-cost devices with poor value and tend to cover a smaller portion of the total bill for a procedure. As a result, healthcare providers have become more inclined to acquire remanufactured medical devices, which can cost end users 50– 75 percent less than a new device.[27] The slow recovery from the 2008–9 economic recession has been a more recent factor affecting sales of remanufactured medical devices. [28] The domestic market for remanufactured medical devices is countercyclical, with recessions strongly correlated with higher demand for these goods. When economic growth slows, many patients—who may lack or lose insurance coverage—commonly postpone healthcare treatments and decline elective procedures, for which imaging equipment is often used. [29] As a result, healthcare providers receive fewer reimbursements for imaging procedures from patients' insurers and find it more difficult to amortize their equipment. Should the equipment need replacing, this trend motivates the end users to lease or buy less-costly remanufactured devices.[30]

U.S. Market for Remanufactured Medical Devices

Market Size

The U.S. market for remanufactured medical devices is estimated to have increased by about 13 percent from $964 million in 2009 to $1.1 billion in 2011 (table 7.2). The U.S. market for remanufactured medical devices is the world's largest, and is estimated to account for more than 35 percent of the global market. [31] U.S. medical device remanufacturers are estimated to export about 35 percent of production, but domestic shipments still account for 90 percent of the U.S. market (table 7.2).

U.S. Buyers and Demand Factors

Hospitals (especially hospitals of 200 beds or less) and diagnostic imaging centers are the two leading buyers of remanufactured medical imaging devices in the United States.[32] Hospitals often purchase remanufactured medical devices to supplement existing equipment and accommodate numerous patients cost effectively.[33] Hospitals and surgical centers are also the leading purchasers of reprocessed SUDs. The principal factor driving demand for remanufactured medical devices is price, as healthcare facilities are in constant need of cost-effective solutions to remain competitive.[34] Healthcare facilities must weigh the cost of acquiring imaging devices (new or remanufactured) with the potential benefits in the form of new patient referrals from physicians and clinicians, along with the potential for increased reimbursements from using the device.[35] Further, healthcare facilities must determine whether their most commonly performed procedures require the use of the most novel technologies, or if a remanufactured device will prove sufficient. Standard imaging equipment or radiation oncology devices, for instance, feature technologies that do not tend to change quickly and may more commonly be purchased remanufactured.[36]

Table 7.2. Medical devices: U.S. market (apparent consumption) for remanufactured goods, 2009–11

Item	2009	2010	2011	% change, 2009–11
	Thousand $			
Production	1,307,588	1,367,739	1,463,313	11.9
Imports	109,776	85,036	110,705	0.8
Exports	453,770	470,858	488,008	7.5
U.S. apparent consumption	963,594	981,917	1,086,010	12.7

Source: USITC staff calculations of weighted responses to the Commission questionnaire.
Note: Totals may not sum due to rounding.

U.S. Sellers and Supply Factors

In the United States, the principal sellers of remanufactured medical devices are OEMs and independent remanufacturers. OEMs sell the majority of their remanufactured medical devices in the U.S. market; however, foreign markets reportedly account for a significant portion of their sales. For instance, foreign markets, including Brazil, India, Italy, Korea, Japan, Portugal, and Spain, account for 45 percent of GE Healthcare's sales of remanufactured medical devices.[37] The principal sources of medical device cores are leading healthcare providers, such as hospitals and imaging centers, and OEMs, which may reacquire the core upon completion of the device's first life cycle.[38] The United States is home to many of the world's largest medical device OEMs and is also a leading market for new medical devices; these two facts translate into a relatively high availability of domestically sourced cores.[39] The supply chains for remanufactured medical devices are closely connected to OEM production and distribution networks. For remanufactured imaging devices, OEMs most commonly acquire cores from their end users upon completion of a lease. In contrast, third-party, or independent, imaging device remanufacturers frequently leverage their relationships with either OEMs or hospitals and imaging centers to acquire cores. For instance, some independent remanufacturers can become "preferred vendors" of OEMs, affording them access to the OEMs' imaging device cores and associated parts.[40] Some third-party remanufacturers may supply regional markets within the United States, sourcing medical device cores from local healthcare facilities and returning the remanufactured product to these end users. Trade-ins by customers who are upgrading to a newer system are estimated to account for almost three-quarters of imaging device cores.[41] With respect to SUDs, cores are most frequently supplied by hospitals and surgical centers. Once remanufactured (reprocessed), these devices are repackaged and returned to the same end users.[42] Two recent trends—hospital closures and the consolidation of imaging centers as hospitals acquire them—have increased the availability of medical device cores.[43] Both developments are believed to be attributable, at least in part, to the Patient Protection and Affordable Care Act (2010), [44] which has reportedly incentivized larger health care companies to seek greater efficiencies.[45]

Estimates of U.S. Trade in Remanufactured Medical Devices and Cores

U.S. Trade in Remanufactured Medical Devices

Exports of Remanufactured Medical Devices

The United States is a net exporter of remanufactured medical devices. U.S. exports are estimated to have increased slightly from $454 million in 2009 to $488 million in 2011 (table 7.3). During the same period, U.S. exports of remanufactured medical devices are estimated to have accounted for 1–2 percent of total U.S. exports of all medical devices (new and remanufactured) over the same period ($35.7 billion in 2011).[46] The EU was the largest market for U.S. exports in 2011, followed by Canada and Brazil. U.S. exports of remanufactured medical devices to FTA partner countries are estimated to have accounted for approximately 20 percent of all U.S. exports of remanufactured medical devices (table 7.4). Canada, Singapore, and Mexico were the largest FTA markets for U.S. exports of remanufactured medical devices in 2011. U.S. exports of remanufactured medical devices by foreign remanufacturers invested in the United States are estimated to have accounted for approximately 13 percent of all U.S. exports of remanufactured medical devices (table 7.5).

Table 7.3. Medical devices: U.S. exports of remanufactured goods by leading destination, 2009–11

Country or market	2009	2010	2011	% change, 2009–11
		Thousand $		
EU	85,655	44,692	47,932	-44.0
Canada	51,252	29,920	39,300	-23.3
Brazil	27,251	[a]33,958	21,968	-19.4
All other countries or markets	[a]289,612	[a]362,289	[a]378,808	30.8
Total	453,770	470,858	488,008	7.5

Source: USITC staff calculations of weighted responses to the Commission questionnaire.
Note: Totals may not sum due to rounding.
[a]Low-precision estimate; RSE below 65 percent.

Imports of Remanufactured Medical Devices

As seen earlier in table 7.2, U.S. imports of remanufactured medical devices are estimated to have remained at about $110 million in both 2009 and 2011; they are estimated to have accounted for less than 1 percent of total U.S. imports of all medical devices (new and remanufactured) during the period ($39.5 billion in 2011).[47] The EU was the leading supplier of U.S. imports, followed by Mexico and Canada (table 7.6). U.S. imports of remanufactured medical devices by foreign remanufacturers invested in the United States are estimated to have accounted for almost 30 percent ($31.2 million) of total U.S. imports of remanufactured medical devices. All such imports came from the EU.

Table 7.4. Medical devices: U.S. exports of remanufactured goods to FTA partners, 2009–11

FTA partner	2009	2010	2011	% change, 2009–11
		Thousand $		
Canada	51,252	29,920	39,300	-23.3
Singapore	7,951	[a]6,305	19,000	139.0
Mexico	15,329	17,507	18,897	23.3
All other FTA partners	27,980	13,004	24,462	-12.6
Total FTA partner exports	102,513	66,735	101,659	-0.7

Source: USITC staff calculations of weighted responses to the Commission questionnaire. Note: Totals may not sum due to rounding.

[a]Low-precision estimate; RSE below 65 percent.

Table 7.5. Medical devices: U.S. exports of remanufactured goods by foreign remanufacturers invested in the United States by leading destination, 2009–11

Country or market	2009	2010	2011	% change, 2009–11
		Thousand $		
EU	27,761	20,352	19,951	-28.1
All other countries or markets	51,413	50,664	45,619	-11.3
Total	79,175	71,015	65,569	-17.2

Source: USITC staff calculations of weighted responses to the Commission questionnaire.

Note: Totals may not sum due to rounding.

Table 7.6. Medical devices: U.S. imports of remanufactured goods by leading source, 2011

Country or market	2011	% share
	Thousand $	
EU	[a]49,127	44.4
Mexico	([b])	([b])
Canada	([b])	([b])
All other countries or markets	16,769	15.1
Total	110,705	100.0

Source: USITC staff calculations of weighted responses to the Commission questionnaire.

Note: Figures for U.S. imports from "all other countries or markets" are calculated. Totals may not sum due to rounding.

[a]Low-precision estimate; RSE below 65 percent. [b]Figures suppressed to prevent the disclosure of CBI.

Trade in Medical Device Cores

Exports of Medical Device Cores

Between 2009 and 2011, the United States switched from being a net exporter to being a net importer of medical device cores. U.S. exports of medical device cores are estimated to have fallen 9 percent from $97 million in 2009 to $88 million in 2011 (figure 7.2), with almost 60 percent ($51 million) of these exports going to FTA partners (figure 7.3). U.S. exports by foreign remanufacturers invested in the United States are estimated to account for almost 20 percent ($16 million) of total U.S. exports of medical device cores, with most likely exported to the EU. U.S. exports are considered to have been limited, in part, by barriers to trade imposed by various countries, the high transportation costs of shipping overseas, the size of cores, and the risks of contamination or damage to key parts while in transit. For example, imaging device cores feature delicate instrumentation that can easily be damaged in transit and are very cumbersome to ship.[48] Instead, the majority of OEMs reportedly remanufacture in the market in which they operate.

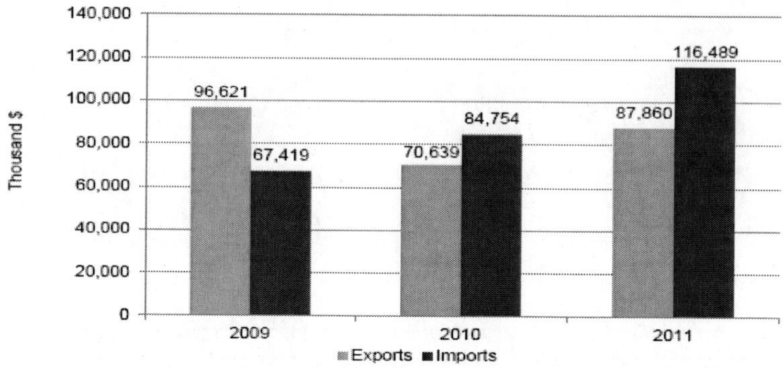

Source: USITC staff calculations of weighted responses to the Commission questionnaire.

Figure 7.2. Medical devices: U.S. core exports and imports, 2009–11.

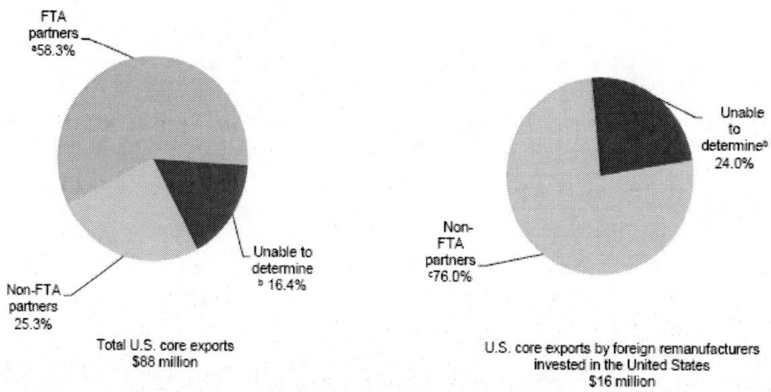

Source: USITC staff calculations of weighted responses to the Commission questionnaire.

Note: USITC staff was unable to estimate U.S. exports of cores by leading destination because of the low statistical significance of the weighted responses to the Commission questionnaire. Instead, USITC staff estimated U.S. exports of cores to FTA and non-FTA partners as two separate groups, which had higher statistical significance.

[a]Low-precision estimate; RSE below 90 percent.

[b]Questionnaire respondents were asked to report the value and destination of their core exports. In some instances, respondents reported the value, but not the destination. In these cases, USITC staff was unable to determine the destination of U.S. core exports. [c]Low-precision estimate; RSE below 80 percent.

Figure 7.3. Medical devices: U.S. core exports to FTA versus non-FTA partners, 2011.

Imports of Medical Device Cores

Nearly 80 percent of the cores consumed in the United States are reportedly sourced domestically, while the remaining 20 percent are imported.[49]

Of these imported cores, the majority are believed to be sourced by third-party providers, who commonly import from the EU and Mexico.[50] U.S. imports of medical device cores are estimated to have increased 73 percent from $67 million in 2009 to $116 million in 2011.

The EU is thought to be among the leading suppliers of medical device cores to the U.S. market (table 7.7). Data for U.S. core imports by foreign remanufacturers invested in the United States are not available.[51]

Table 7.7. Medical devices: U.S. core imports from FTA versus non-FTA countries, 2011

Leading source	2011	% share
	Thousand $	
FTA countries	[a]1,491	1.3
Non-FTA countries	[b]91,897	78.9
Unable to determinec	23,101	19.8
Total	[b]116,489	100.0

Source: USITC staff calculations of weighted responses to the Commission questionnaire.

Note: USITC staff was unable to estimate U.S. imports of cores by leading source because of the low statistical significance of the weighted responses to the Commission questionnaire. Instead, USITC staff estimated U.S. imports of cores from FTA and non-FTA partners as two separate groups, which had higher statistical significance. Totals may not sum due to rounding.

[a]Low-precision estimate; RSE below 65 percent.

[b]Low-precision estimate; RSE below 80 percent.

[c]Questionnaire respondents were asked to report the value and source of their core imports. In some instances, respondents reported the value, but not the source. In these cases, USITC staff was unable to determine the source of U.S. core imports.

Barriers to Trade in Remanufactured Medical Devices

Remanufactured medical devices are subject to numerous regulations that vary across countries. According to the U.S. Department of Commerce, roughly 16 countries place restrictions on imports of remanufactured medical devices, while 5 countries impose outright bans.[52] While this means that most countries are open to these imports, [53] many of those that do prohibit or restrict them are potentially large markets. For instance, Brazil places restrictions on imports of remanufactured medical devices, while China prohibits imports. [54] Brazil also announced a proposal in 2011 to prohibit imports of remanufactured medical imaging equipment that competes with local production.[55]

Medical Imaging Equipment

Although most countries permit unrestricted imports of remanufactured medical imaging equipment, these imports may be subject to extensive regulatory requirements and government procurement policies. As in the United States, products with a greater perceived risk to patients are subject to more rigorous safety requirements in the EU. In addition, government procurement policies in 24 countries discourage or prevent publicly run healthcare institutions from importing remanufactured medical equipment and commonly favor new technologies.[56] Although privately run facilities in these countries are often permitted to source remanufactured devices, the market presence of these institutions is relatively limited.

Single-use Devices

Export markets for remanufactured SUDs are limited, in part, because of fragmented regulations. Canada and the EU are believed to be the largest markets for remanufactured SUDs. However, both markets lack a comprehensive federal regulatory framework, which can be a significant barrier for U.S. remanufacturers of SUDs seeking to export.[57] In the EU, for instance, each of the 27 member states has different policies for reprocessing, while Canada has deferred to the provinces for regulation.[58]

FDI in the Medical Devices Remanufacturing Sector

FDI flows in U.S. medical device remanufacturing are estimated to have increased modestly from $11.4 million in 2009 to $12 million in 2011 (table 7.8). This figure is likely substantially underestimated, as witnesses at the Commission hearing estimated that approximately $180 million is invested annually in remanufacturing of medical imaging devices by multinational firms, with much of this funding directed towards plant improvements, capital equipment, and labor. [59] FDI in the U.S. medical device remanufacturing sector is estimated to have supported around 400 full-time employees in this period, or almost 10 percent of total remanufacturing employment (about 4,100 employees) in this sector.

Table 7.8. Medical devices: FDI and associated employment in U.S. remanufacturing, 2009–11

Item	2009	2010	2011	% change, 2009–11
	Thousand $			
Outbound FDI	[a]6,577	[a]7,608	[b]10,992	67.1
Inbound FDI	[a]11,361	[a]13,085	[a]11,962	5.3
	Full-time workers			
U.S. employment associated with inbound FDI	410	395	375	–8.5

Source: USITC staff calculations of weighted responses to the Commission questionnaire.
[a]Low-precision estimate; RSE below 80 percent.
[b]Low-precision estimate; RSE below 65 percent.

The United States has long been regarded as the world's leading healthcare market for both new and remanufactured medical equipment. Each of the leading foreign remanufacturers of medical devices and equipment perform remanufacturing operations in the United States, including Netherlands-based Philips in Bothell, WA, and Cleveland, OH;[60] German-based Siemens in Hockehim, IL;[61] and Japan-based Toshiba in Irvine, CA.[62]

In an effort to penetrate leading foreign healthcare markets, many of the largest multinational OEMs—both U.S. based and foreign—have invested in remanufacturing facilities abroad. While more mature regional markets in the EU have traditionally attracted leading OEMs such as Philips, [63] companies are also increasingly establishing remanufacturing facilities in developing markets. For instance, GE Healthcare and Siemens have both invested in India.[64] This strategy may be employed to penetrate these markets while complying with local laws, which in some cases may prohibit the importation of remanufactured medical devices, but allow devices that are remanufactured in their country to be sold domestically.[65] For instance, China prohibits imports of remanufactured medical devices, but allows their domestic production. Hence, GE Healthcare has established a facility in Beijing to remanufacture CT, MRI, X-ray, and nuclear devices.[66] Likewise, in Brazil, GE invested $50 million in a facility that will both manufacture and remanufacture various types of diagnostic imaging equipment, including positron emission tomography, CT, MRI, and monitoring systems.[67]

References

Association of Medical Device Reprocessors (AMDR). Written testimony submitted to the U.S. International Trade Commission in connection with inv. no 332-525, *Remanufactured Goods: An Overview of the U.S. and Global Industries, Markets, and Trade*, February 28, 2012.

Barlow, Rick Dana. “Shaking Off a Stigma.” *Outpatient Care Technology,* April/May 2006. http://www.ardusmedical.com/pdf/O406 Refurb SF.pdf.

———. “How Second-hand Equipment Can Mean First-rate Buys.” *ITN*, June 11, 2007. http://www.itnonline.com/article/how-second-hand-equipment-can-mean-first-rate-buys.

Barnes, Julie. “Moving Away from Fee-for-Service.” *The Atlantic,* May 7, 2012. http://www.theatlantic.com/health/archive/2012/05/moving-away-from-fee-for-service/256755/.

Beinke, Hans. Siemens Medical Solutions. Written testimony submitted to the U.S. International Trade Commission in connection with inv. no 332-525, *Remanufactured Goods: An Overview of the U.S. and Global Industries, Markets, and Trade*, February 9, 2012.

Berry, Emily. “Mergers between Insurers and Hospitals Expected to Accelerate.” *American Medical News,* October 25, 2010. http://www.ama-assn.org/amednews/2010 /10/25/bisa1025.htm.

COCIR. *Good Refurbishment Practice for Medical Imaging Equipment.* White Paper, October 2009. http://www.cocir.org/content.php?level1 =9&mode =9.

Dineen, John. “Healthymagination Investor Update.” PowerPoint presentation, GE Healthcare, November 9, 2010. http://www.ge.com/pdf/investors /events/11092010 /ge_healthcare_analyst_meeting_presentation_ 11092010.pdf.

Dworkin, Neil. “Health Costs: Behind the Numbers.” Letter to the Editor. *New York Times*, May 6, 2012. http://www.nytimes.com/2012/05/07 /opinion/health-costs-behind-the-numbers.html?_r=1.

Elario, David. “Remanufactured Policy Roundtable.” PowerPoint presentation, Washington, DC, January 24, 2011.

GE Healthcare. “GE Healthcare Opens First Plant in Brazil,” July 22, 2010. http://newsroom.gehealthcare.com/articles/ge-healthcare-opens-first-plant-in-brazil/.

———. “GoldSeal Refurbished Systems.” http://www.gehealthcare.com/euen /pre-owned-equipment/index.html (accessed August 30, 2012).

Gibson, Diane. "Business Trends in Shipping Logistics for Medical Equipment." *Medical Device and Diagnostic Industry,* February 2, 2012. http://www.mddionline.com /article/trends-shippinglogistics.

IAMERS. *Responsible Handling of Pre-owned Medical Diagnostic Imaging Equipment.* White paper, n.d. http://98.130.199.220/downloads/White Paper5149.pdf (accessed May 6, 2012).

IBISWorld. *Medical Device Manufacturing in the US.* IBISWorld Industry Report 33451b, June 2011.

Imaging Technology News (ITN). "Importance of Refurbished Equipment to be Highlighted at ASTRO," November 1, 2010. http://www.itnonline.com /article/importance-refurbished-equipment-behighlighted-astro.

———. "Toshiba Introduces Toshiba Assurance Refurbishment Program," November 10, 2009. http://www.itnonline.com/article/toshiba-introduces-toshiba-assurance-refurbishment-program.

Levin Associates. "Health Care M&A Spending Rises 11% in 2011." Press release, January 25, 2012. http://www.levinassociates.com/pr2012 /pr1201mamq4.

Martin, Anne B., David Lassman, Benjamin Washington, Aaron Catlin, and the National Health Expenditure Accounts Team. "Growth in U.S. Health Spending Remained Slow in 2010; Health Share of Gross Domestic Product Was Unchanged from 2009." *Health Affairs* 31, no. 1 (2012).

Medical Devices Today. "Reimbursement Execs Assess Challenges for '09," January 5, 2008. http://www.medicaldevicestoday.com/2009/01 /reim bursement-execs-assess-challenges-for09.html.

Delporte, Christopher, Michael Barbella, and Niki Arrowsmith, eds. "The Top 30." *Medical Product Outsourcing* 9, no. 6 (July/August 2011).

"Lawmaker Proposes Repealing Controversial Device Tax." *Medical Product Outsourcing* 8, no. 4 (May 2010).

Philips. "Refurbished Equipment: The Diamond Select Program," n.d. http://www. healthcare.philips.com/main/products/refurbished_systems/#
(accessed August 31, 2012).

PRWeb. "The Market Outlook for Refurbished Medical Devices to 2016," January 16, 2012.

———. "Global Medical Imaging Equipment Services Market to Reach US$12.35 Billion by 2017, According to New Report by Global Industry Analysts, Inc.," April 5, 2012.

Pyrek, Kelly M. "Reprocessing of Medical Devices: Government Intensifies its Scrutiny As Clinicians Debate Patient-Safety Issues," February 1, 2003. http://www.surgistrategies. com/articles/2003/02/reprocessing-of-medical-devices-116124.aspx.

Siemens. "Proven Excellence," n.d. http://usa.siemens.com/proven-excellence (accessed November 15, 2011).

Singh, Seema. "GE Sets Up Facility in Bangalore to Refurbish Medical Equipment," May 18, 2009. http://www.livemint.com/Companies /MhVHvtR7Mh1aC0IFoHvmdO/GE-sets-up-facility-inBangalore-to-refurbish-medical-equipm.html.

Standard & Poor's (S&P). *Industry Surveys Healthcare: Products and Supplies*, February 4, 2010. SterilMed. "FAQs," n.d. http://www.sterilmed.com /rg-faq.shtml (accessed May 16, 2012).

U.S. Department of Commerce (USDOC). International Trade Administration (ITA). *Global Import Regulations for Pre-Owned (Used and Refurbished) Medical Devices.* 6th ed.,

May 2008. http://www.ita.doc.gov/td/health /PreOwnedMedEquipment FINAL 050608.pdf.

U.S. Food and Drug Administration (FDA). "Who Must Register, List and Pay a Fee," updated August 10, 2012. http://www.fda.gov/MedicalDevices /DeviceRegulationand Guidance/HowtoMarketYourDevice/R egistrationandListing/ucm053165.htm (accessed August 23, 2012).

U.S. International Trade Commission (USITC). Hearing transcript in connection with inv. no. 332-525, *Remanufactured Goods: An Overview of the U.S. and Global Industries, Markets, and Trade,* February 28, 2012.

World Health Organization (WHO). Medical Device Regulations: Global Overview and Guiding Principles. Geneva: WHO, 2003.

RETREADED TIRES

Description of Tire Retreading Activities

Tire retreading is a process of applying a new tread to an inspected used or worn tire ("casing"), effectively restoring it to a like-new condition.[1] Retreaded tires are typically used on a wide variety of vehicles, including all types of trucks and buses, heavy construction and agricultural equipment, aircraft, and passenger vehicles. The retreading process consists of several steps, including inspecting and repairing the casing, buffing the worn tread away, and bonding or curing a new tread to the tire body under specific heat, time, and pressure conditions. Retreaded tires produced from used tire casings cost significantly less than new tires. In addition, tires can be retreaded multiple times from a given tire casing, thereby conserving energy and raw materials and providing an environmentally friendlier alternative to the disposal of used tire casings.

Size of Retreading in the United States

Industry Structure

In 2011, approximately 680 firms reportedly produced retreaded truck tires, 18 firms produced off-the-road (OTR) retreaded tires,[2] five firms produced aircraft retreads, and only four smaller shops produced passenger vehicle retreads.[3] Altogether, these retread shops vary in size from small "mom and pop" operations producing 20 or fewer retreaded tires per day to very large shops processing more than 1,000 retreaded tires daily.[4] In recent years, the U.S. retreaded tire industry has undergone consolidation, resulting in fewer but larger firms led by the "Big Three" multinational tire manufacturers and retreaders—Goodyear (U.S. ownership), Bridgestone (Japanese ownership), and Michelin (French ownership).[5] Approximately 90 percent of all retreaded tires produced in the United States are estimated to fall under the umbrella of the Big Three tire producers.[6] The Big Three firms operate their own wholly owned retreading operations, as well as myriad downstream licensed franchises that use authorized production equipment and that obtain their rubber-based feedstock materials from the Big Three firms. Firms that retread tires produce few, if any, new tires in the same facilities, although an estimated 85 percent of U.S. tire retreaders also sell new tires.[7] Goodyear retread operations are organized under the Wingfoot/Goodyear banner (brand), Bridgestone under the Bandag banner, and Michelin under the Michelin banner. Other major U.S. tire retreaders include Marangoni (Italian ownership), which has a widespread dealership network across the United States; Oliver (Michelin); and Continental (German ownership).[8]

Production, Investment, and Employment

U.S. production of retreaded tires is estimated to have increased by 35 percent from $1.0 billion in 2009 to $1.4 billion in 2011 (table 8.1). Despite this growth, retreaded tires are estimated to have accounted for only about 3 percent of total sales by U.S. firms within the tires sector during the period ($48 billion in 2011). [9] Investment in tire retreading operations is estimated to have increased by about 20 percent from $20 million in 2009 to $24 million in 2011. Industry consolidation has reportedly spurred larger investments in U.S. retreading operations, running into the millions of dollars. [10] U.S. employment in tire retreading is estimated to have increased as well, growing 24 percent from about 3,900 in 2009 to 4,900 in 2011.[11]

Table 8.1. Retreaded tires: Remanufacturing production, investment, and employment, 2009–11

Item	2009	2010	2011	% change, 2009–11
	Thousand $			
Production	1,038,679	1,188,315	1,399,088	34.7
Investment	19,894	29,109	23,874	20.0
	Full-time workers			
Employment	3,935	4,368	4,880	24.0

Source: USITC staff calculations of weighted responses to the Commission questionnaire.

A rebounding U.S. economy and a relatively short supply of new tires, due in part to increasing transportation demand, helped to raise demand for and production of retreaded tires, especially for the truck, aircraft, and OTR tire segments. In 2011, U.S. tire retreaders operated at near full capacity, limited principally by the availability of used tire casings. [12] Between 2009 and 2011, an estimated 60 percent of U.S. tire retreaders expanded production capacity,[13] and the U.S. retreaded tire industry is anticipated to continue to upgrade and expand retreading operations in response to continued growth of the U.S. economy. [14]

U.S. Market for Retreaded Tires

Market Size

The U.S. market for retreaded tires is estimated to have increased by 35 percent from $1.0 billion in 2009 to $1.4 billion in 2011 (table 8.2). Both exports and imports of retreaded tires are minimal relative to domestic production. Historically, the United States has experienced a balance between the production and consumption of retreaded tires, and the vast majority of available tire casings are recycled back into downstream retreading operations for domestic consumption.[15]

Table 8.2. Retreaded tires: U.S. market (apparent consumption) for retreaded tires, 2009–11

Item	2009	2010	2011	% change, 2009–11
		Thousand $		
Production	1,038,679	1,188,315	1,399,088	34.7
Imports	6,014	8,513	11,446	90.3
Exports	15,904	16,495	18,545	16.6
U.S. apparent consumption	1,028,789	1,180,334	1,391,989	35.3

Sources: Estimates of production based on USITC staff calculations of weighted responses to the Commission questionnaire. Trade data from official statistics of the USDOC for HTS subheadings 4012.11, 4012.12, 4012.13, and 4012.19.

Note: Figures for U.S. apparent consumption are calculated.

U.S. Buyers and Demand Factors

Demand for tires (whether new or retreaded) depends on the general health of the domestic economy and on demand for goods and services. Transportation demand is a major factor in the truck, bus, aviation, and passenger car tire industries, while demand for construction, mining, and agricultural equipment is a driver for the OTR tire segment. Demand for retreaded tires relative to new tires is driven by a number of factors, including competitive quality and cost savings, the availability of casings and end-use applications, and government regulations, including those that encourage or require the use of retreaded tires.[16] Consumers in both the public and private sectors depend on a reliable supply of competitively priced retreaded tires, which are generally equal in quality to new tires, but cost only about one-half as much.[17]

Before the 2008–09 economic recession, buoyant U.S. demand for goods and services and transportation services contributed to $1.6 billion in retreaded tire shipments in 2007.[18] Following the recession, there was a gradual recovery in the U.S. economy, leading to increased transportation demand and U.S. shipments of retreaded tires. In 2010 and 2011, increasing demand for retreaded tires in the trucking, OTR, aviation, and agricultural sectors reportedly led to a shortage of used tire casings.[19] During the same time period, domestic shortages in new replacement tires for trucks, coupled with global supply shortages of both casings and new tires, reportedly pushed up demand for retreaded tires, prompting some tire retreaders to ask truck fleets to extend their retreading cycle—the number of times a worn casing could be recycled—to higher limits. [20] (Many commercial trucking fleets plan their new tire purchases with the intention of having their worn casings retreaded two or more times as a routine part of their tire budgets.)[21]

The tractor-trailer (18-wheeler semi-rig) market is reported to be the largest consumer of retreaded tires in the United States, although the OTR and aviation sectors are also reportedly large.[22] In 2011, retreaded tires accounted for 47 percent of U.S. truck tire consumption (new and retreaded). Bridgestone (Bandag), Goodyear, and Michelin collectively accounted for 93 percent of the retreaded truck tire market that year. [23] Despite a shortage of truck tire casings in 2011, U.S. tire retreaders reportedly produced 15.5 million retreaded truck tires, 7 percent more than in 2010. The average reported price of a retreaded truck tire, including the cost of the casing, was $248, or 60 percent of the cost of a new truck tire ($416).[24] By using retreaded tires instead of new tires, the truck industry reportedly saves more than $3 billion annually.[25]

OTR tires are typically larger, heavier tires used on construction, mining, and agricultural equipment. These tires, by virtue of their size and their rubber, steel, and other materials content, are more expensive to produce than typical truck or bus retreads; they are priced accordingly as specialty retreads. Retreaded OTR tires are usually produced in low volumes by specialty shops, and command a price premium compared with other types of retreaded tires. Still, large OTR tire retreads, such as for earth-movers, may cost less than one-third as much as a comparable new OTR tire.[26]

All commercial airlines, as well as the military, rely on retreaded aircraft tires. In fact, about 80 percent of all aircraft tires now in service in the United States are retreaded. By using retreaded tires, the commercial and military aircraft sectors reportedly save more than $100 million annually.[27] The Big Three tire producers and Desser Aircraft Tire, an independent producer, are the principal aviation retread manufacturers in the United States.[28]

U.S. Sellers and Supply Factors

Retreaded tires are sold, serviced, and warrantied in essentially the same way as new tires.[29] Direct retail sales are estimated to account for 70 percent of retreaded tire sales, while sales to wholesalers account for the remainder.[30] Typically, large commercial trucking and bus fleets, aviation firms, OTR operations, and federal agencies negotiate contracts with the downstream distributors of retread operations to handle the supply, replacement, repair, and recycling of retreaded tires. As noted earlier, the majority of retread tire distributors are wholly owned or linked through franchise agreements to the major U.S. retreaders.[31]

The major U.S. commercial truck tire producers typically guarantee retread customers that spent tire casings can be retreaded up to three times, and tire retreaders offer additional economic incentives, such as warranties, if their particular tire brands are used.[32] Additionally, retread distributors provide warranties on retreaded tires based on mileage, time on the road, and tread wear, among other considerations.[33]

In spite of a shortage of tire casings in 2011, several U.S. retreaders, including the Big Three tire producers, Marangoni, Continental, and other downstream dealer franchises and independent retreaders, continued to expand production capacity at existing plants and planned new plant expansions to keep pace with increasing demand.[34] In January 2012, for example, Marangoni announced a $10 million expansion project at its tread rubber facility at Madison, TN, to provide retread stock to its multi-state retreading dealers.[35] The Big Three tire producers also announced expansion plans, as did H&H Industries Inc. (Oak Hill, OH), one of the largest independent OTR tire retreaders in the country, which planned to increase its export business.[36]

Continental opened a new tread rubber plant in Mexico, and franchised an expanding retread operation in the United States.[37] In addition, certain retreaders have opted to switch to producing other brand-name retreaded tires, reflecting the intense competition among domestic producers.[38]

Estimates of Trade in Retreaded Tires and Casings

U.S. and Global Trade in Retreaded Tires

U.S. Trade in Retreaded Tires

The United States is a small net exporter of retreaded tires. Between 2009 and 2011, U.S. exports of retreaded tires increased by about 17 percent to $18.5 million (table 8.3),[39] but accounted for less than 1 percent of total tire exports (new and retreaded) ($5.1 billion in 2011).[40] Canada and Mexico accounted for 36 percent of U.S. exports of retreaded tires in 2011. FTA partners together accounted for 52 percent ($9.6 million) of total U.S. exports of retreaded tires in 2011, down from 58 percent ($9.2 million) in 2009 (table 8.4).

Table 8.3. Retreaded tires: U.S. exports of retreaded tires by leading destination, 2009–11

Country or market	2009	2010	2011	% change, 2009–11
	Thousand $			
Mexico	3,929	3,697	4,353	10.8
Canada	3,942	3,642	2,255	–42.8
Vietnam	1,981	3,299	1,809	–8.7
South Africa	210	298	1,701	710.0
EU	1,697	654	1,315	–22.5
Guatemala	293	413	1,044	256.3
All other countries or markets	3,853	4,492	6,068	57.5
Total	15,904	16,495	18,545	16.6

Source: Official statistics of the USDOC, HTS subheadings 4012.11, 4012.12, 4012.13 and 4012.19.
Note: Totals may not sum due to rounding.

Table 8.4. Retreaded tires: U.S. exports of retreaded tires to FTA partners, 2009–11

FTA partner	2009	2010	2011	% change, 2009–11
	Thousand $			
Mexico	3,929	3,697	4,353	10.8
CAFTA-DRa	1,007	1,213	2,679	166.0
Canada	3,942	3,642	2,255	–42.8
Australia	130	236	150	15.4
Israel	0	66	109	n/a
Singapore	16	6	66	312.5
Chile	119	281	28	–76.5
Other FTA partners[b]	12	63	0	n/a
Total	9,155	9,204	9,641	5.3

Source: Official statistics of the USDOC, HTS subheadings 4012.11, 4012.12, 4012.13 and 4012.19.
Note: Totals may not sum due to rounding.
N/a = not applicable.
[a]Dominican Republican-Central America-United States Free Trade Agreement partners.
[b]Jordan, Morocco, and Peru.

Partner countries to the North America Free Trade Agreement (NAFTA) and the Dominican Republic-Central America-United States Free Trade Agreement (CAFTADR) accounted for practically all of U.S. retreaded tire exports to FTA partner countries. Between 2009 and 2011, U.S. imports of retreaded tires jumped by 90 percent to $11.4 million, primarily from Canada (table 8.5), but accounted for less than 1 percent of total tire imports (new and retreaded) during the same period ($13.2 billion in 2011).[41]

Table 8.5. Retreaded tires: U.S. imports of retreaded tires by leading source, 2009–11

Country or market	2009	2010	2011	% change, 2009–11
		Thousand $		
Canada	3,389	4,802	7,612	124.6
EU	613	1,041	1,327	116.5
Korea	161	218	1,064	560.9
China	365	1,335	561	53.7
Mexico	8	127	378	4,625.0
Colombia	0	62	217	n/a
Israel	5	47	114	2,180.0
All other countries or markets	1,474	882	171	–88.4
Total	6,015	8,514	11,444	90.3

Source: Official statistics of the USDOC, HTS subheadings 4012.11, 4012.12, 4012.13 and 4012.19.
Note: Totals may not sum due to rounding. N/a = not applicable.

Global Trade in Retreaded Tires

Global exports of retreaded tires reached $253 million in 2011, up from $187 million in 2009 (table 8.6). The EU accounted for almost one-half of global exports in 2011. In contrast, the United States accounted for less than 10 percent. Russia is the largest importer of retreaded tires, accounting for 12 percent of global imports in 2011 (table 8.7). The United States accounted for 7 percent of global imports of retreaded tires in 2011.

Table 8.6. Retreaded tires: Global exports of retreaded tires by leading exporters, 2009–11

Country or market	2009	2010	2011	% change, 2009–11
		Thousand $		
EU	91,608	108,817	123,387	34.7
Thailand	16,013	19,497	21,583	34.8
Hong Kong	14,265	21,437	21,206	48.7
South Korea	11,129	13,813	20,694	85.9
United States	15,341	17,085	18,909	23.3
China	6,179	10,961	13,592	120.0
Canada	5,586	6,805	8,669	55.2
Malaysia	2,875	2,815	3,837	33.5
Australia	1,515	1,554	3,165	108.9
Singapore	1,496	964	2,684	79.4
All other countries or markets	20,650	16,089	15,552	–24.7
Total	186,657	219,838	253,278	35.7

Source: GTIS, Harmonized Commodity Description and Coding System (HS) subheadings 4012.11, 4012.12, 4012.12 and 4012.19.

Note: Totals may not sum due to rounding.

U.S. and Global Trade in Tire Casings

U.S. Trade in Tire Casings

Between 2009 and 2011, U.S. exports of tire casings increased 34 percent to $39.3 million (table 8.8). U.S. imports of tire casings increased by 80 percent to $43.9 million during the same period (table 8.9), probably due to the shortage of domestic truck and OTR tire casings noted previously. Japan and the EU supplied 65 percent of U.S. imports of tire casings in 2011, likely reflecting the cross-border trade flows of Bridgestone, Goodyear, and Michelin, three major retreaded tire producers in the U.S. market.

Table 8.7. Retreaded tires: Global imports of retreaded tires by leading importers, 2009-11

Country or market	2009	2010	2011	% change, 2009–11
	Thousand $			
Russia	8,397	13,075	17,877	112.9
Switzerland	13,441	14,622	17,474	30.0
EU	8,735	11,026	13,915	59.3
United States	6,013	8,513	11,444	90.3
Singapore	5,403	5,579	9,405	74.1
Mexico	8,448	8,019	8,840	4.6
Malaysia	5,526	5,543	6,759	22.3
Canada	5,349	7,124	5,519	3.2
Ukraine	2,134	6,272	5,506	158.0
Norway	3,013	3,585	5,447	80.8
All other countries or markets	37,529	42,531	52,343	39.5
Total	103,989	125,890	154,528	48.6

Source: GTIS, HS subheadings 4012.11, 4012.12, 4012.13 and 4012.19.
Note: Totals may not sum due to rounding.

Table 8.8. Retreaded tires: U.S. exports of used pneumatic tires (including tire casings)[a] by leading destination, 2009–11

Country or market	2009	2010	2011	% change, 2009–11
	Thousand $			
Mexico	11,617	12,541	13,868	19.4
Guatemala	2,777	3,629	4,471	61.0
Vietnam	222	211	4,401	1,882.4
Canada	2,241	2,343	2,957	32.0
Honduras	1,035	1,194	1,810	74.9
Dominican Republic	954	1,326	1,352	41.7
EU	5,792	5,094	1,293	-77.7
All other countries or markets	4620	6656	9135	97.7
Total	29,258	32,994	39,287	34.3

Source: Official statistics of the USDOC, HTS subheadings 4012.20. Note: Totals may not sum due to rounding.

[a]Used pneumatic tires is a basket category that includes tire casings suitable for retreading as well as other used tires that may not be suitable for retreading. U.S. totals may be different from those shown in table 8.10 (global exports of used pneumatic tires), due to differences in reporting or to data suppression to safeguard CBI.

Table 8.9. Retreaded tires: U.S. imports of used pneumatic tires (including tire casings)[a] by leading source, 2009-11

Country or market	2009	2010	2011	% change, 2009–11
	Thousand $			
Japan	9,988	12,888	17,347	73.7
EU	9,541	8,962	11,276	18.2
Canada	1,861	2,947	5,627	202.4
Korea	303	965	2,466	713.9
Thailand	1,210	884	2,145	77.3
China	736	1,143	1,812	146.2
Australia	72	233	1,543	2,043.1
All other countries or markets	719	1,150	1,643	128.5
Total	24,430	29,172	43,859	79.5

Source: Official statistics of the USDOC, HTS subheading 4012.20. Note: Totals may not sum due to rounding.

[a]Used pneumatic tires is a basket category that includes tire casings suitable for retreading as well as other used tires that may not be suitable for retreading. U.S. totals may be different from those shown in table 8.11 (global imports of used pneumatic tires) due to differences in reporting or to data suppression to safeguard CBI.

Global Trade in Tire Casings

Between 2009 and 2011, global exports of tire casings increased by 27 percent to $383.5 million (table 8.10). Japan and the EU collectively accounted for over two-thirds of global exports. In contrast, the United States was the largest importer of tire casings, although it never accounted for more than 15 percent of global imports of casings during the period (table 8.11).

Table 8.10. Retreaded tires: Global exports of used pneumatic tires (including tire casings)[a] by leading exporters, 2009–11

Country or market	2009	2010	2011	% change, 2009–11
		Thousand $		
Japan	117,668	129,352	154,200	31.0
EU	89,323	92,866	105,118	17.7
United States	31,090	35,350	44,186	42.1
Canada	6,912	10,191	16,011	131.6
Switzerland	10,945	12,050	15,175	38.6
Hong Kong	6,297	11,412	9,014	43.1
Korea	3,308	3,486	7,148	116.1
Thailand	3,602	2,765	5,670	57.4
Singapore	4,515	5,346	4,425	-2.0
Australia	5,476	5,001	3,808	-30.5
All other countries or markets	22,358	19,020	18,722	-16.3
Total	301,494	326,839	383,477	27.2

Source: GTIS, HS subheading 4012.20.

Note: Totals may not sum due to rounding.

[a]Used pneumatic tires is a basket category that includes tire casings suitable for retreading as well as other used tires that are not suitable for retreading.

FDI in the Retreaded Tires Sector

No figures were available for inbound FDI in tire retreading activities in the United States and associated employment, as well as for U.S. investment in foreign tire retreading activities.[42] However, as previously noted, multinational tire manufacturers and retreaders with foreign ownership play a prominent role in the U.S. retreaded tire industry, and these firms likely account for a substantial amount of inbound FDI and employment in U.S. retreading activities. Leading sources of inbound FDI likely include Japan (home of Bridgestone), France (Michelin), Italy (Marangoni), and Germany (Continental). Likewise, U.S.-based tire retreader Goodyear operates tire retreading facilities in Europe, Latin America, and the Asia-Pacific, and likely accounts for a substantial amount of outbound FDI for foreign retreading activities in those destinations.[43]

Table 8.11. Retreaded tires: Global imports of used pneumatic tires (including tire casings)[a] by leading importers, 2009–11

Country or market	2009	2010	2011	% change, 2009–11
		Thousand $		
United States	24,543	29,171	43,869	78.7
EU	35,671	34,785	38,086	6.8
Mexico	20,049	22,530	27,994	39.6
Thailand	21,702	25,908	24,679	13.7
Ghana	([b])	12,759	16,738	n/a
Guatemala	8,360	11,493	15,076	80.3
South Africa	8,226	9,467	12,968	57.6
Georgia	([b])	10,588	12,157	n/a
Hong Kong	13,410	11,833	12,056	–10.1
Korea	4,889	7,851	10,488	114.5
All other countries or markets	75,366	78,805	85,480	13.4
Total	212,216	255,190	299,591	41.2

Source: GTIS, HS subheading 4012.20.

Note: Totals may not sum due to rounding. N/a = not applicable.

[a]Used pneumatic tires is a basket category that includes tire casings suitable for retreading as well as other used tires that may not be suitable for retreading.

[b]None reported.

References

Bridgestone Bandag Tire Solutions. "Bridgestone Bandag Tire Solutions Announces Campaign to Encourage Fleets to Retread Instead." News release, March 21, 2011.

Continental. "First ContiLifeCycle Dealership Opens in U.S." Press release, September 23, 2011.

———. "Shipments Begin from Continental's Flat Tread Production Facility in Mexico." Press release, May 26, 2011.

Goodyear Tire & Rubber Company. *Form 10-K Annual Report,* 2011.

Marangoni Tread North America. "Marangoni Expands North American Operations." Press release, August 31, 2011.

Modern Tire Dealer, "The Top 100 Retreaders in the U.S.," April 2011.

———. "2011 Was a Great Year for Retreading," January 2012.

———. "Everyone Scrambles for Usable Truck Tires," January 2012.

———. "Prepping for Growth," April 2012.

Rubber Manufacturers Association. "U.S. Scrap Tire Management Summary 2005–2009," October 2011.

Tire Industry Association (TIA) and Tire Retread & Repair Information Bureau (TRIB). "Remanufactured Goods: An Overview of the U.S. and Global Industries, Markets and Trade." *Retread Tires—The First Industrial Recycling Program*, February 27, 2012.

———. "Understanding Retreading," 2009.

Ulrich, Bob. "A Case for Casings." *Modern Tire Dealer*, April 2011.

U.S. Department of Commerce (USDOC). Census Bureau (Census). *2007 Economic Census*, "Tire Retreading." http://factfinder2.census.gov /faces/nav/jsf/pages/index.xhtml (August 26, 2012).

U.S. International Trade Commission (USITC). Hearing transcript in connection with inv. no. 332-525, *Remanufactured Goods: An Overview of the U.S. and Global Industries, Markets, and Trade,* February 28, 2012.

OTHER REMANUFACTURING SECTORS

Introduction

This chapter provides an overview of U.S. remanufacturing activities in six other remanufacturing sectors that vary in size: consumer products, electrical apparatus, locomotives, machinery, office furniture, and restaurant equipment.[1]

The scope of remanufacturing activities varies significantly across these sectors. For instance, as noted in chapter 2, although machinery is estimated to be the fourth-largest remanufacturing sector (after aerospace, HDOR equipment, and motor vehicle parts), it comprises a diverse range of products and cuts across a variety of industries, each affected by its own combination of demand and supply factors.

By contrast, the locomotive sector is concentrated among a few industry players and covers a relatively limited range of products that are remanufactured in small volumes.

For each of these six sectors, this chapter summarizes information on the size and scope of U.S. remanufacturing activities, U.S. trade in remanufactured goods and related cores, and FDI in remanufacturing. Where more information is available, this chapter provides specific sector detail, such as for consumer products and locomotives.

Overview of Sector Remanufacturing Activities

Industry structure varies among the six sectors included in this chapter, and so do the types of firms that participate in the various sectors (table 9.1).

Between 2009 and 2011, the combined production of remanufactured goods in these six sectors is estimated to have increased by 25 percent, from $8.4 billion in 2009 to $10.4 billion in 2011 (table 9.2). Nevertheless, production of remanufactured goods in these sectors is estimated to have accounted for less than 1 percent of the combined total sales by U.S. firms within these sectors during 2009–11 ($1.4 trillion in 2011). Between 2009 and 2011, domestic investment in remanufacturing activities in these sectors is estimated to have increased by 141 percent, from $325 million in 2009 to $783 million in 2011. Most of this increase was driven by investment in the machinery sector, likely reflecting investment by one or two firms in remanufacturing facilities during 2009–11. On the other hand, during the same period, remanufacturing employment in these sectors is estimated to have increased only modestly, from about 54,500 full-time workers in 2009 to 57,500 in 2011.

Table 9.1. Other remanufacturing sectors: Industry overview

Sector	Products typically subject to remanufacturing	Industry structure	Type of firms
Consumer products	Certain consumer electronics and household appliances, most notably cell phones	Fragmented— hundreds to thousands of small companies; only 5–10 companies likely to employ 100 or more workers	Independent remanufacturers, contract remanufacturers, OEMs, and some U.S. distributors
Electrical apparatus	Primarily electric motors, switchgears, and transformers; also relays and industrial controls	Fragmented—500– 1,000 firms	Independent remanufacturers (typically small ones); some OEMs that produce new electrical apparatus
Locomotives	Locomotive engines, drive motors, axles, other subsystems	Highly concentrated— 6 OEMs[a]	OEMs that produce new locomotives
Machinery	Industrial valves, turbines, machine tools, textile machinery, compressors	Fragmented—an estimated 1,000–2,000 firms	Repair shops that both repair and remanufacture; OEMs that produce new machinery
Office furniture	Office systems furniture, laminated work surfaces, reupholstered office seating	Somewhat concentrated—roughly 50 firms	Mostly independent remanufacturers; some OEMs that produce new furniture
Restaurant equipment	Ovens, beverage dispensers, food preparation tables	Somewhat concentrated; mostly SMEs as well as some larger U.S. distributors	Independent remanufacturers; some distributors

Source: Compiled by USITC staff.

[a]Brookville Equipment Co., Caterpillar/Progress Rail Service, GE/Rail Products Division (a unit of GE Transportation), MotivePower-WABTEC, National Railway Equipment Co., and RJ Corman/Railpower.

U.S. Market for Remanufactured Goods

Market Size

The size of the U.S. market for remanufactured goods in these sectors varies considerably (table 9.3). Not surprisingly, the U.S. market for remanufactured machinery is the largest among these six sectors; its apparent consumption is nearly five times that of consumer products, the next largest sector by this measure. The Commission did not estimate the size of the remanufactured restaurant equipment sector separately; however, industry sources indicate that its total U.S. market size is likely to have amounted to $500 million in 2011.[2] The degree to which remanufactured goods in these sectors are imported also varies substantially. For instance, imports are estimated to have accounted for 36 percent of the U.S.

market for remanufactured consumer products, but for only 6 percent of the U.S. market for remanufactured machinery.

Table 9.2. Other remanufacturing sectors: Remanufacturing production, investment, and employment, combined sectors, 2009–11

Item	2009	2010	2011	% change, 2009–11
	Thousand $			
Production				
Machinery	4,059,570	4,774,291	5,795,105	42.8
Consumer products	557,612	567,320	659,175	18.2
Other[a]	3,745,608	3,499,468	3,973,923	6.1
Total production	8,362,790	8,841,078	10,428,203	24.7
Investment				
Machinery	[b]206,579	[c]554,371	[c]711,008	244.2
Consumer products	[d]34,035	7,416	4,948	-85.5
Other[a]	84,021	67,333	67,537	-19.6
Total investment	324,635	[c]629,119	[b]783,493	141.3
		Full-time workers		
Employment				
Machinery	24,814	24,870	26,843	8.2
Consumer products	8,216	7,733	7,613	-7.3
Other[a]	21,467	22,210	22,999	7.1
Total employment	54,497	54,814	57,455	5.4

Source: USITC staff calculations of weighted responses to the Commission questionnaire. Note: Totals may not sum due to rounding.

[a]Includes the electrical apparatus, locomotive systems, office furniture, and restaurant equipment remanufacturing sectors.

[b]Low-precision estimate; RSE below 80 percent.

[c]Low-precision estimate; RSE below 100 percent.

[d]Low-precision estimate; RSE below 65 percent.

U.S. Market Participants and Market Factors

U.S. Buyers and Demand Factors

In the United States, buyers of remanufactured goods within these six sectors are predominantly the final end users of the products.

Factors affecting the demand for remanufactured goods in these sectors are most often (1) the price of remanufactured goods compared to new ones, and (2) consumer preferences (table 9.4).

Other important demand factors are the lead times for obtaining remanufactured goods versus new ones, and price competition from alternative products.

U.S. Sellers and Supply Factors

Remanufacturers are among the leading sellers in the United States of remanufactured goods within these sectors, although wholesalers play a role for some products as well.

Much as they were for other remanufacturing sectors, the leading factors influencing the supply of remanufactured goods within the six sectors covered in this chapter are the availability and price of cores (table 9.5). Hence, supply chain management is a key component of competitiveness in all of these sectors.

Table 9.3. Other remanufacturing sectors: U.S. market (apparent consumption) for remanufactured goods by sector, 2009–11

Item	2009	2010	2011	% change, 2009–11
		Thousand $		
Production				
Machinery	4,059,570	4,774,291	5,795,105	42.8
Consumer products	557,612	567,320	659,175	18.2
Other[a]	3,745,608	3,499,468	3,973,923	6.1
Total	8,362,790	8,841,078	10,428,203	24.7
U.S. imports				
Machinery	[b]136,460	[b]504,861	[b]268,256	96.6
Consumer products	[c]325,095	[d]394,162	[c]360,264	10.8
Other[a]	39,977	41,688	40,683	1.8
Total	501,532	[d]940,711	669,203	33.4
U.S. exports				
Machinery	[d]858,281	884,716	1,348,734	57.1
Consumer products	[c]13,733	[d]16,608	21,151	54.0
Other[a]	128,672	135,613	224,627	74.6
Total	1,000,686	1,036,937	1,594,512	59.3
U.S. apparent consumption				
Machinery	3,337,749	4,394,436	4,714,627	41.3
Consumer products	868,974	944,874	998,288	14.9
Other[a]	3,656,913	3,405,543	3,789,979	3.6
Total	7,863,636	8,744,852	9,502,894	20.8

Source: USITC staff calculations of weighted responses to the Commission questionnaire. *Note*: Totals may not sum due to rounding.

[a]Includes the electrical apparatus, locomotive systems, office furniture, and restaurant equipment remanufacturing sectors.

[b]Low-precision estimate; RSE below 100 percent.

[c]Low-precision estimate; RSE below 80 percent.

[d]Low-precision estimate; RSE below 65 percent.

Other supply factors include the cost of production (including cores) of remanufactured goods versus new products, and the suitability of certain products for remanufacturing. For example, many consumer electronics may not be economically viable to remanufacture, due to rapid advancements in technology and changing consumer preferences.

For cell phones, for instance, the cost of remanufactured goods versus new ones, changing technology, consumer preferences, and product design influence whether or not they are remanufactured rather than refurbished (box 9.1).

Table 9.4. Other remanufacturing sectors: U.S. buyers and demand factors

Sector	U.S. buyers	Principal factors influencing demand
Consumer products	End users	Consumer preferences and price.
Electrical apparatus	End users of equipment—firms in the mining, manufacturing, and electrical utility sectors	Price and lead times of new products. Price of remanufactured goods can be 30–50 percent less than for new goods; lead times for remanufactured goods can be up to 50 percent shorter than for new goods.
Locomotives	Rail operators	Economic activity, as it affects shipments of goods by rail; competition from alternative forms of transportation (e.g., trucking).
Machinery	End users; wholesalers that sell to end users	Price and lead times of new products. Price of remanufactured goods can be 30–50 percent less than for new goods; lead times for remanufactured goods can be up to 50 percent shorter than for new goods.
Office furniture	SMEs; start-up businesses; schools; nonprofit institutions	Consumer preferences and price. In addition, commercial construction activity drives demand for both new and remanufactured office furniture.
Restaurant equipment	SMEs in the restaurant or institutional food service markets (e.g., hotels, bars, cafeterias, restaurants)	Price of remanufactured goods compared to new goods.

Sources: Compiled by USITC staff; industry representative, email message to USITC staff, August 20, 2012; Electric Control Equipment Co., "Why Would Companies Choose to Buy Remanufactured Gear?" March 22, 2011; GE Energy, "Transformer Remanufacturing Fact Sheet," January 2010; Association of American Railroads, "Class I Railroad Statistics," June 17, 20111, 15–16; Foran, "Locomotive Rebuilds," December 10, 2009.

Table 9.5. Other remanufacturing sectors: U.S. sellers and supply factors

Sector	U.S. sellers	Principal supply factors
Consumer products	Remanufacturers; retailers	Cost of remanufacturing (versus refurbishing); technology and design of the product (see box 9.1).
Electrical apparatus	OEMs; independent remanufacturers; smaller resellers; wholesalers	Core availability and price—determined by amount of (1) old equipment from industrial plant closings or modernizations; and (2) end users subcontracting to have specific products remanufactured.
Locomotives	OEMs that remanufacture their own locomotives	High cost of technology (including research, design, development, and testing), which limits firms' ability to enter the market as a remanufacturer.
Machinery	OEMs; independent remanufacturers; wholesalers for certain products	Core availability and price. Cores are obtained from customers upgrading machinery, including trade-ins of older equipment; from auctions; and from customers seeking to have a specific product remanufactured.
Office furniture	OEMs that remanufacture their own products; independent remanufacturers, including those that also refurbish (i.e., cosmetically enhance in this case); wholesalers; retailers	Core availability and price. Increased OEM sales of new office furniture in 2011 may increase availability of used work stations and panels to be remanufactured.
Restaurant equipment	Independent remanufacturers; regional equipment distributors; wholesalers that sell to restaurants and other institutional food equipment suppliers	Core availability. "Take back" arrangements with regional food distributors provide discounts on remanufactured appliances in exchange for old equipment.

Sources: Compiled by USITC staff; industry representative, email message to USITC staff, January 17, 2012; industry representative, interview by USITC staff, February 22, 2012; Malavi, Train, Subway, and Transit Car Manufacturing in the US, May 2011, 7, 25.

Box 9.1. Refurbishing versus remanufacturing of consumer goods: The example of cell phones

Consumer products comprise a broad range of goods, including consumer electronics (e.g., cell phones and DVD players) and appliances (e.g., clothes dryers and vacuum cleaners). For the most part, consumer goods tend to be refurbished rather than remanufactured (rebuilt from cores).

To see why, it is important to understand the difference between the two processes. The refurbisher of a consumer good tends to test it, possibly make minor repairs, clean the good for cosmetic purposes, and then package it for resale. In contrast, a remanufacturer will dismantle the good, replace any worn parts, and reassemble and test it before packaging it for resale.

The most important reason that consumer products are more apt to be refurbished than remanufactured is that most "used" consumer products are actually goods that have been returned to a retailer because the buyer changed his or her mind, and not because the good in question is at the end of its useful life.

As a result, the good may only require a cosmetic refurbishing before it can be resold. There are several other reasons remanufacturing is less frequently used for consumer products: (1) the cost of a remanufactured good could be too close to the cost of a new good to justify the effort; (2) the technology, particularly for consumer electronics, is constantly changing, and older technology is often considerably less desirable to consumers than new technology; (3) the "form factor" (size, shape, and configuration) of the older good is less desirable than that of the new good; and (4) the goods were not designed and assembled with future remanufacturing in mind. Roughly speaking, used consumer electronics with a value of less than $100 are often sold "as is"; those with a value of $100–$300 may be either refurbished or sold as is; and those valued at more than $300 may be either refurbished or remanufactured.

Unlike many other consumer products, cell phones are often not used up to the point where they have reached the end of their useful lives.

A cell phone user will frequently discard a perfectly usable cell phone in order to acquire a new one with more and better features that has been offered free or at a low price by the service provider as an inducement to sign a new service contract. The service provider can then either resell the used cell phone outright or offer it as a warranty replacement after cosmetic refurbishing or remanufacturing.

In deciding what to do with used cell phones, an additional consideration is a difference in wireless technologies.

Code division multiple access (CDMA) technology is used principally in the United States and some parts of Asia, while global system for mobile communications (GSM) technology is used in the EU and most other countries. As a result, most cell phones that are recycled in the United States are suited for use only in the United States; only a minimal number can be exported.

Source: Industry representative, interview by USITC staff, St. Louis, MO, October 21, 2011.

Estimates of U.S. Trade in Remanufactured Goods and Cores

Trade in Remanufactured Goods

Exports of Remanufactured Goods

The United States is a net exporter of remanufactured goods in these sectors on a combined basis. As noted previously in table 9.3, combined U.S. exports of remanufactured goods in these sectors are estimated to have increased by about 60 percent during the study period, from $1.0 billion in 2009 to $1.6 billion in 2011. Nevertheless, these U.S. exports are estimated to have accounted for less than 1 percent of total U.S. exports of all manufactured products (new and remanufactured) in these sectors during 2009–11 ($195.7 billion in 2011).[3] Again, remanufactured machinery dominates in this category: U.S. exports of these goods are estimated to have accounted for 85 percent ($1.3 billion) of total U.S. exports of remanufactured goods within these six sectors in 2011, and likely include high-value, low-volume products, such as turbines.

Export markets for these goods were similar to those for other remanufactured products. NAFTA partners Mexico and Canada, as well as the EU, were the leading markets for U.S. exports of remanufactured goods in these sectors during 2009–11 (table 9.6). U.S. exports to FTA partners are estimated to have accounted for 50 percent of total U.S. exports of remanufactured goods in these sectors in 2011. Mexico alone is estimated to have accounted for over 50 percent ($422 million) of U.S. exports to FTA partners in 2011.

Table 9.6. Other remanufacturing sectors: U.S. exports of remanufactured goods by leading destination, combined sectors, 2009–11

Country or market	2009	2010	2011	% change, 2009–11
		Thousand $		
Leading destination:				
Mexico	[a]393,182	[b]139,964	[a]422,240	7.4
EU	[a]126,611	[b]121,729	[b]190,528	50.5
Canada	90,359	106,762	150,089	66.1
All other countries or markets	390,534	668,482	831,656	113.0
Total	1,000,686	1,036,937	1,594,512	59.3
To FTA partners:				
Mexico	[a]393,182	[b]139,964	[a]422,240	7.4
Canada	90,359	106,762	150,089	66.1
CAFTA-DR[c]	[b]55,135	25,869	[a]82,440	49.5
All other FTA partners	35,389	56,187	140,147	296.0
Total	[a]574,065	328,782	[d]794,916	38.5

Source: USITC staff calculations of weighted responses to the Commission questionnaire.

Note: Figures for exports to "all other countries or markets" and to "all other FTA partners" are calculated. Totals may not sum due to rounding.

[a]Low-precision estimate; RSE below 100 percent. [b]Low-precision estimate; RSE below 80 percent.

[c]Dominican Republic-Central America-United States Free Trade Agreement partners.

[d]Low-precision estimate; RSE below 65 percent.

U.S. exports of remanufactured goods in these sectors by foreign remanufacturers invested in the United States are estimated to have increased by 53 percent, from $735 million in 2009 to $1.1 billion in 2011, accounting for 70 percent of total U.S. exports of remanufactured goods in these sectors in 2011 (table 9.7). Most of these types of U.S. exports were machinery, and principal export markets included Mexico and the EU.

Table 9.7. Other remanufacturing sectors: U.S. exports of remanufactured goods by foreign remanufacturers invested in the United States by sector and by leading destination, 2009–11

Item	2009	2010	2011	% change, 2009–11
		Thousand $		
Sector:				
Machinery	[a]732,464	667,682	[a]1,120,608	53.0
Other[b]	[c]2,294	[c]2,106	[c]2,013	-12.2
Total	[a]734,758	669,788	[a]1,122,621	52.8
Leading destination:				
Mexico	[d]375,332	[d]120,667	[d]399,398	6.4
EU	[d]115,513	[d]107,832	[d]142,150	23.1
CAFTA-DR[e]	[d]40,845	[c]14,749	[d]71,502	75.1
All other countries or markets	203,069	426,541	509,571	150.9
Total	[a]734,758	669,788	[a]1,122,621	52.8

Source: USITC staff calculations of weighted responses to the Commission questionnaire.

Note: There were no reported U.S. exports of remanufactured consumer products by foreign remanufacturers invested in the United States. Figures for U.S. exports to "all other countries or markets" are calculated. Totals may not sum due to rounding.

[a]Low-precision estimate; RSE below 65 percent.

[b]Includes the electrical apparatus, locomotive systems, office furniture, and restaurant equipment remanufacturing sectors.

[c]Low-precision estimate; RSE below 80 percent.

[d]Low-precision estimate; RSE below 100 percent.

[e]Dominican Republic-Central America-United States Free Trade Agreement partners.

Imports of Remanufactured Goods

As noted previously in table 9.3, U.S. imports of remanufactured goods in these sectors are estimated to have increased by 33 percent, from $502 million in 2009 to $669 million 2011. However, combined U.S. imports of remanufactured goods in these sectors are estimated to have accounted for less than 1 percent of total U.S. imports of all manufactured products (new and remanufactured) in these sectors during 2009–11 ($250.0 billion in 2011). [4] U.S. imports of consumer products (primarily consumer electronics) are estimated to have accounted for more than one-half ($360 million) of total U.S. imports of remanufactured goods in these sectors in 2011. China and Mexico were the principal suppliers of imported remanufactured goods in these sectors to the U.S. market (figure 9.1).

U.S. imports of remanufactured machinery by foreign remanufacturers invested in the United States are estimated to have almost doubled, from $135 million in 2009 to $263 million in 2011, accounting for 39 percent of total U.S. imports by foreign remanufacturers invested in the United States in these sectors in 2011 (table 9.8). The substantial increase in these types of U.S. imports likely reflects imports of low-volume, high-value machinery products, such as turbines. The EU was reported to be an important supplier of remanufactured machinery to the U.S. market.

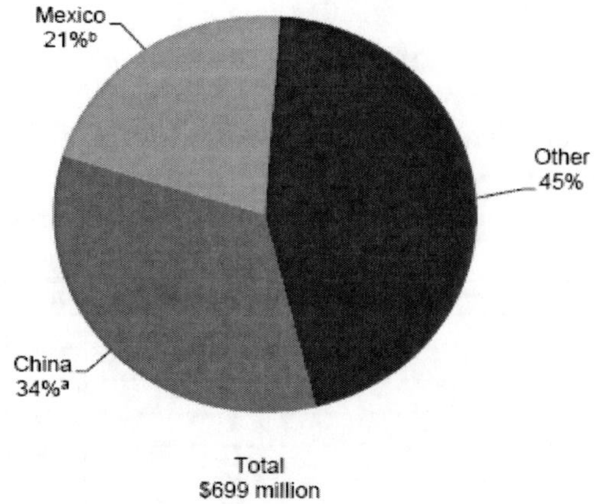

Source: USITC staff calculations of weighted responses to the Commission.
[a]Low-precision estimate; RSE below 100 percent. [b]Low-precision estimate; RSE below 65 percent.

Figure 9.1. Other remanufacturing sectors: U.S. imports of remanufactured goods by leading source, combined sectors, 2011.

Trade in Cores

Exports of Cores

The United States is a net exporter of cores for the six sectors covered within this chapter. Between 2009 and 2011, U.S. core exports are estimated to have more than quadrupled, from $67 million in 2009 to $317 million in 2011 (table 9.9). The majority of these were destined for China (figure 9.2). Oftentimes, exported cores are remanufactured abroad and imported back into the United States as a finished product. In one example, Solar Turbines, a subsidiary of Caterpillar Inc., remanufactures industrial gas turbines in an overhaul facility in Desoto, TX. Cores and other parts are sent to the company's Turbotec Remanufacturing Facility near Tijuana, Mexico, where the components and parts are remanufactured and returned to the facility in Texas for use in remanufacturing.[5]

Table 9.8. Other remanufacturing sectors: U.S. imports of remanufactured goods by foreign remanufacturers invested in the United States by sector, 2009–11

Sector	2009	2010	2011	% change, 2009–11
		Thousand $		
Machinery	[a]135,066	[a]500,795	[a]263,182	94.9
Consumer products	([b])	([b])	([b])	([b])
Other[c]	([d])	([d])	([d])	([d])
Total	135,066	500,795	263,182	94.9

Source: USITC staff calculations of weighted responses to the Commission questionnaire.
[a]Low-precision estimate; RSE below 100 percent.
[b]Figures suppressed to prevent the disclosure of CBI.
[c]The electrical apparatus, locomotive systems, office furniture, and restaurant equipment remanufacturing sectors. [d]None reported.

Table 9.9. Other remanufacturing sectors: U.S. core exports by sector, 2009–11

Sector	2009	2010	2011	% change, 2009–11
		Thousand $		
Machinery	[a]52,646	[b]81,312	[b]115,949	120.2
Consumer products	[b]14,405	[b]67,429	[c]198,322	1,276.8
Other[d]	[b]45	[b]1,640	[b]2,738	5,984.4
Total	67,096	150,381	[a]317,009	372.5

Source: USITC staff calculations of weighted responses to the Commission questionnaire.

Note: Figures for U.S. core exports to "all other countries" are calculated. Totals may not sum due to rounding.

[a]Low-precision estimate; RSE below 65 percent.

[b]Low-precision estimate; RSE below 80 percent.

[c]Low-precision estimate; RSE below 100 percent.

[d]Includes the electrical apparatus, locomotive systems, office furniture, and restaurant equipment remanufacturing sectors.

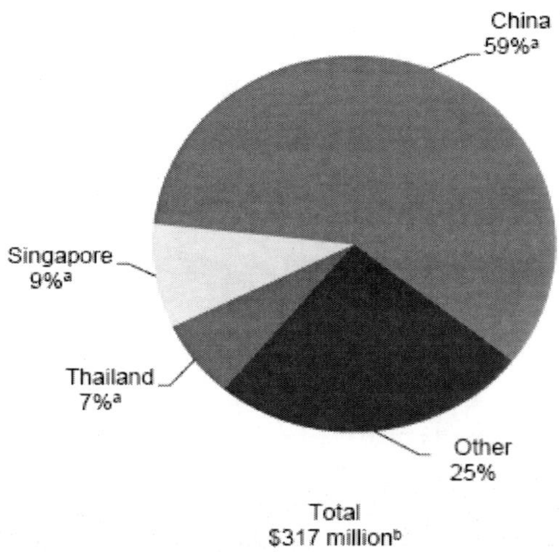

Source: USITC staff calculations of weighted responses to the Commission questionnaire.
[a]Low-precision estimate; RSE below 100 percent. bLow-precision estimate; RSE below 65 percent.

Figure 9.2. Other remanufacturing sectors: U.S. core exports by leading destination, combined sectors, 2011.

U.S. core exports by foreign remanufacturers invested in the United States are estimated to have more than doubled, from $48 million in 2009 to $100 million in 2011, accounting for 32 percent of total U.S. core exports from these sectors in 2011 (table 9.10). The vast majority of these exports were machinery cores.

Imports of Cores

Between 2009 and 2011, U.S. core imports in these six sectors are estimated to have increased more than 10-fold to $132 million, the vast majority of which were machinery cores (table 9.11). U.S. core imports in these sectors by foreign remanufacturers invested in the United States are estimated to have accounted for 90 percent ($119 million) of total U.S. core imports relating to these six sectors in 2011.

Table 9.10. Other remanufacturing sectors: U.S. core exports by foreign remanufacturers invested in the United States by sector, 2009–11

Sector	2009	2010	2011	% change, 2009–11
		Thousand $		
Machinery	[a]47,570	[a]70,931	[a]100,207	110.7
Consumer products	([b])	([b])	([b])	
Other[c]		([d])	([d])	
Total	47,570	70,931	100,207	110.7

Source: USITC staff calculations of weighted responses to the Commission questionnaire.
[a]Low-precision estimate; RSE below 80 percent.
[b]Figures suppressed to prevent the disclosure of CBI.
[c]Includes the electrical apparatus, locomotive systems, office furniture, and restaurant equipment remanufacturing sectors.
[d]None reported.

Table 9.11. Other remanufacturing sectors: U.S. core imports by sector and by foreign remanufacturers invested in the United States, 2009–11

Item	2009	2010	2011	% change, 2009–11
		Thousand $		
Total U.S. imports by sector				
Machinery	10,335	[a]14,263	[b]122,205	1,082.4
Consumer products	([c])	([c])	([c])	n/a
Other[d]	[a]19,640	[a]23,268	10,026	–49.0
Total	29,975	37,531	[e]132,231	341.1
U.S. imports by foreign remanufacturers invested in the United States by sector				
Machinery	9,386	[a]12,538	[b]118,934	1,167.1
Consumer products	([c])	([c])	([c])	n/a
Other[d]	([c])	([c])	([c])	n/a
Total	9,386	12,538	118,934	1,167.1

Source: USITC staff calculations of weighted responses to the Commission questionnaire. Note: Totals may not sum due to rounding. N/a = not applicable.
[a]Low-precision estimate; RSE below 65 percent.
[b]Low-precision estimate; RSE below 100 percent.
[c]None reported.
[d]Includes the electrical apparatus, locomotive systems, office furniture, and restaurant equipment remanufacturing sectors.
[e]Low-precision estimate; RSE below 80 percent.

FDI in Other Remanufacturing Sectors

FDI flows in U.S. remanufacturing in these six sectors are estimated to have increased sharply between 2009 and 2011, and likely reflect U.S. acquisitions or other large investments in U.S. remanufacturing operations in the machinery sector by one or more foreign firms during this period (table 9.12). U.S. remanufacturing employment supported by FDI is estimated to have accounted for 20 percent (about 11,300 full-time workers) of total U.S. remanufacturing employment in these sectors during the period, primarily in the machinery remanufacturing sector.

Table 9.12. Other remanufacturing sectors: FDI and associated employment in U.S. remanufacturing by sector, 2009–11

Item	2009	2010	2011	% change, 2009–11
	Thousand $			
Inbound FDI by sector				
Machinery	[a]22,955	[b]414,512	[b]377,901	1,546.3
Other[c]	[a]7,024	[a]10,690	[a]11,380	62.0
Total inbound FDI	[d]29,979	[b]425,202	[b]389,282	1,198.5
	Full-time workers			
U.S. employment				
Machinery	10,102	10,262	11,077	9.7
Other[c]	[d]263	[d]267	271	3.0
Total U.S. employment	10,366	10,529	11,348	9.5

Source: USITC staff calculations of weighted responses to the Commission questionnaire. Note: Totals may not sum due to rounding. N/a = not applicable.

[a]Low-precision estimate; RSE below 80 percent.

[b]Low-precision estimate; RSE below 100 percent.

[c]Includes the consumer products, electrical apparatus, locomotive systems, office furniture, and restaurant equipment remanufacturing sectors.

[d]Low-precision estimate; RSE below 65 percent.

Table 9.13. Other remanufacturing sectors: U.S. outbound FDI in foreign remanufacturing by sector, 2009–11

Item	2009	2010	2011	% change, 2009–11
		Thousand $		
Outbound FDI by sector				
Machinery	5,080	2,213	4,397	–13.4
Consumer products	[a]1,766	[b]955	599	–66.1
Other[c]	20,872	10,103	1,421	–93.2
Total outbound FDI	27,718	13,272	6,417	–76.8
Outbound FDI by destination				
Mexico	3,698	2,914	2,241	–39.4
Canada	[b]2,186	[a]618	[b]1,626	–25.6
Malaysia	[b]12,993	[b]4,855	[b]788	–93.9
All other countries	8,841	4,885	1,761	–80.1
Total outbound FDI	27,718	13,272	6,417	–76.8

Source: USITC staff calculations of weighted responses to the Commission questionnaire.

Note: Figures for U.S. outbound FDI to "all other countries" are calculated. Totals may not sum due to rounding.

[a]Low-precision estimate; RSE below 80 percent. [b]Low-precision estimate; RSE below 65 percent.

[c]Includes the electrical apparatus, locomotive systems, office furniture, and restaurant equipment remanufacturing sectors.

U.S. outbound FDI flows in foreign remanufacturing activities were small, and estimated to have declined, from $28 million to $6 million during 2009–11 (table 9.13). More than one-half of U.S. outbound FDI flows were directed to Mexico and Canada.

REFERENCES

Association of American Railroads. "Class I Railroad Statistics," June 17, 2011.

Electric Control Equipment Co., Inc. "Why Would Companies Choose to Buy Remanufactured Gear?" June 19, 2012. http://www.electriccontrol.com /blog/why-would-companies-choose-to-buyremanufactured-gear/.

Foran, Pat. "Locomotive Rebuilds—A Market Update." *Progressive Railroading*, December 10, 2009. http://www.progressiverailroading.com /pr/article.asp?id=22112.

GE Energy. "Transformer Remanufacturing Fact Sheet," June 2012.

Malavi, Justin. *Train, Subway & Transit Car Manufacturing in the US*. IBISWorld Industry Report 33651, May 2011.

Solar Turbines, Inc. "Turbotec Remanufacturing Facility," March 2009.

GLOBAL TRADE, INDUSTRIES, AND MARKETS FOR REMANUFACTURED GOODS

Introduction

This chapter provides an overview of global trade, industries, and markets for remanufactured goods, including government measures that affect remanufacturers in selected countries. [1] Statistical data on global trade in remanufactured goods are largely unavailable. However, the United States and the EU likely account for the bulk of such trade, as most of the remanufacturing-intensive sectors described in preceding chapters of this report are well established in both economies.

Although repair activities are fairly common in a diverse range of industry sectors throughout the world, remanufacturing as a distinct industry is a relatively recent development in many countries, and these countries' trade in remanufactured goods is likely small. Key among the primary factors limiting global trade in these goods is the absence of a commonly accepted legal definition of remanufactured goods. As a consequence, many customs authorities classify remanufactured goods as used products, which are often heavily restricted and in many cases prohibited. Reasons for such restrictions vary, but are often based on environmental, health, and consumer safety considerations. Other reasons include providing competitive advantages to, or protecting, competing domestic industries. Also, there are historical perceptions of low quality and the view in many developing countries that remanufactured goods are "discards" from larger markets.

This chapter describes the principal industries and markets for remanufactured goods in selected Asia-Pacific Economic Cooperation (APEC) Forum and G-20 economies that are thought to account for the majority of global remanufacturing activity.[2] This chapter focuses on the EU (which not only has a more developed remanufacturing sector, but generally is more open to trade in remanufactured goods) and on Brazil, India, and China (which have remanufacturing sectors in varying degrees of development and restrict trade in remanufactured goods). To a lesser extent, information is provided on remanufacturing activities in Singapore and Korea, which have both recognized remanufacturing as part of broader efforts to promote sustainability and reduce waste in industrial activities, as well as on other selected APEC countries for which information is available.

Foreign Remanufacturing Industries and Markets

European Union

Remanufacturing in the EU occurs across a range of industry sectors similar to those in the United States. Aerospace, motor vehicle parts, HDOR equipment, and machinery are the dominant remanufacturing sectors, followed by medical devices and furniture. Remanufacturing occurs throughout the EU, although the United Kingdom (UK) and, to a lesser extent, Germany are thought to account for the bulk of it. For the most part, EU members view remanufacturing as a tool to conserve natural resources, reduce costs for manufacturers and consumers, and protect the environment. Hence, the EU's governing body, the European Commission (EC), has issued several mandatory legal directives to foster the growth of remanufacturing in Europe.[3]

Remanufacturing Industries and Markets

The Commission could find no estimates of the size of remanufacturing in the EU as a whole. A 2004 study estimated remanufactured goods production in the UK, the EU's leading remanufacturer, at £4.9 billion (approximately $7.8 billion) in 2002, and UK remanufacturing employment at 44,300 workers.[4] Aerospace remanufacturing accounted for 40 percent (£2 billion or $3.2 billion) of the total that year, while machinery and motor vehicle parts remanufacturing each accounted for 11 percent (£530 million or $837.4 million).

Aerospace

As noted in chapter 3, safety considerations require the complete overhaul or rotation of aircraft components and assemblies on a regular basis. Only companies approved or certified by the European Aviation Safety Agency (EASA) can remanufacture aircraft components or perform remanufacturing activities in the EU.[5] The leading EU aircraft engine OEMs are Rolls-Royce (EU), followed by SNECMA and its subsidiary Turbomeca (both of France). Whereas all aircraft engine OEMs perform maintenance and repair on aircraft, a few airlines also have their own in-house maintenance and repair operations to overhaul engines and major component systems. These include Lufthansa Technik (Germany) and Air France/KLM Engineering (France/Netherlands).[6] In addition, U.S.-based engine OEM Pratt & Whitney has an aircraft engine maintenance and repair facility in Ireland. These same companies also perform maintenance and repair activities in the United States, as noted in chapter 3.

Motor Vehicle Parts

The value of motor vehicle parts remanufacturing in the EU was estimated at $5.2 billion in 2005.[7] The UK alone was estimated to produce $1.2 billion worth of remanufactured motor vehicle parts in 2009, [8] or a little less than one-fifth the estimated size of U.S. production of remanufactured motor vehicle parts that year.

Remanufacturing activities occur throughout the EU, although remanufacturing of motor vehicle parts is concentrated in the UK and Germany, reflecting the prominence of these countries in motor vehicle production and repair. There are at least 115 motor vehicle parts remanufacturers in the EU, and 18 motor vehicle part core suppliers. [9] Core suppliers are located in the UK, Germany, France, Italy, the Netherlands, Hungary, Spain, and Sweden. At least six U.S.-based OEMs remanufacture motor vehicle parts in the EU as well, including Caterpillar, Cummins, John Deere, Delco Remy, Meritor (formerly Rockwell), and TRW.

The increased availability of relatively inexpensive new aftermarket parts from Asia reportedly has made it difficult for remanufacturers in Western Europe to remain competitive. In response, some European remanufactures have sought to economize by transferring remanufacturing operations to central and eastern European countries that have lower labor costs.

HDOR Equipment

As noted in chapter 4, the HDOR sector consists of construction equipment, including lifting and handling equipment, and off-the-road (OTR) equipment used in the agriculture, construction, and mining industries. The leading HDOR equipment remanufacturers in the EU are OEMs such as Caterpillar (U.S. ownership), JCB (UK ownership), and Komatsu (Japanese ownership). They often contract out remanufacturing of specific parts to smaller, specialized firms.[10] For example, Wealdstone Engineering (UK ownership) in Northampton, UK, performs certain remanufacturing services for Caterpillar.

IT Products

Imaging products, including inkjet and toner cartridges, account for the bulk of the value of remanufacturing in the IT products sector in the EU. According to the European Toner and Inkjet Remanufacturers Association (ETIRA), there are 1,400 European firms that remanufacture inkjet and toner cartridges. [11] The remanufactured cartridge industry encompasses many small firms, but includes several large multinational firms as well.[12] For example, U.S.-owned Xerox reportedly remanufactures large-volume printers and components for a range of printing equipment for the EU market in Mitcheldean, Gloucestershire, UK, and in Dundalk, Ireland. [13] Océ remanufactures printers and scanners in the Netherlands.[14]

ETIRA estimates that the market for remanufactured toner cartridges in Europe is €1.2 billion (around $1.5 billion).[15] Remanufactured cartridges (excluding color laser toner) have a 27 percent share of the EU market for toner cartridges. As noted in chapter 6, the remanufacturing process is labor intensive, as it consists of disassembling spent inkjet and toner cartridges, cleaning the cartridges, refilling them, and testing the cartridges to ensure that they are restored to original working condition. The remanufacturing industry relies on high prices of new ink and toner cartridges to justify the customer's purchase of lower-cost remanufactured cartridges. Some used cartridges are reportedly exported to low-labor-cost countries in Asia for cleaning and refilling and then reimported into the EU for distribution.[16]

Medical Devices

As noted in chapter 7, remanufactured medical devices typically include refurbished medical imaging equipment and reprocessed single-use devices (SUDs). Used medical devices that have been remanufactured are often exported from the EU to customers that cannot afford the expense of new, technologically advanced equipment. Leading medical device remanufacturers in the EU include GE Healthcare (UK), Philips Healthcare (Netherlands), Siemens (Germany), and Toshiba (Switzerland).[17]

Measures Affecting Remanufacturers

For the most part, EC directives encourage the development and growth of remanufacturing industries in Europe. Restrictive policies, such as those relating to electronic equipment and the disposal of hazardous waste when remanufacturing automotive equipment, are intended to protect consumers from inferior quality goods, shield workers from exposure to carcinogens, and safeguard the environment. For instance, the EU's Directive on End-of-Life Vehicles provides guidance for harmonizing end-of-life regulations for motor vehicles among EU member states.[18] The stated goals of the directive are to minimize the impact of end-of-life vehicles on the environment, to conserve energy, and to ensure the smooth operation of the internal market and avoid distortions of competition in the EU. The directive sets targets for reuse, recycling, and recovery. As more EU members comply with the directive and intra-EU standards are established, the directive will likely lead to an expanded use of remanufactured motor vehicle parts in the EU.[19]

Similarly, the EU's Directive on Waste Electrical and Electronic Equipment (WEEE) sets targets for waste recovery, recycling, and reuse for certain electrical and electronic devices, and establishes that producers are responsible for financing the management of waste from their products.[20] The effects of WEEE on remanufacturers likely vary: WEEE may encourage reuse, including remanufacturing, but remanufacturers that are OEMs must also bear the costs of compliance as original producers of covered equipment.

As noted in chapter 6, the EU's Restriction of the Use of Certain Hazardous Substances (RoHS) Directive regulates the use and amount of certain substances, such as lead, in electric and electronic components in the EU market. RoHS is of particular importance to the IT sector because IT products are likely to contain substances that are restricted under the directive. The directive is intended to ensure that used electronic components do not contain substances that may damage human health or the environment. However, these restrictions have reportedly driven up remanufacturing costs for some EU remanufacturers by prohibiting the reuse of components containing these substances.[21]

Brazil

Remanufacturing in Brazil is reportedly a small yet growing industrial activity. Although information on the size of remanufacturing activities in Brazil is largely unavailable, leading Brazilian remanufacturing sectors reportedly include aerospace, motor vehicle parts, HDOR equipment, and IT products (predominantly printer cartridge remanufacturing). The government of Brazil does not distinguish between remanufactured goods and used goods for import declaration purposes. Also, the government imposes a series of restrictions on imports of remanufactured goods and effectively prohibits imports of cores. As a result, Brazil's trade in remanufactured goods is reportedly negligible, and its remanufacturing sector largely serves the domestic market.

Remanufacturing Industries and Markets

Although there is not an official definition of what constitutes remanufacturing in Brazil, the country's general repair industry (which includes remanufacturing) consists of thousands of firms. Most of these companies are small, employing 20 or fewer workers. However, domestic remanufacturers also include OEMs and independent firms. OEMs typically remanufacture and sell products that were originally produced by them. Independent remanufacturers, on the other hand, may either purchase cores on the open market and remanufacture and sell them, or remanufacture cores provided by a specific customer.

Aerospace

Like the maintenance and repair industry in the United States, the Brazilian maintenance and repair industry services aircraft and parts for both domestic and international airlines operating in the country, as well as for OEM producers of aircraft and parts. Five firms reportedly provide the bulk of maintenance and repair services for aircraft in Brazil. U.S.-based multinational firms Rockwell Collins, Goodrich Aerostructures, and Honeywell are among the leading maintenance and repair organizations (MROs) in Brazil.

Several trends have led to rapid increases in demand for maintenance and repair services in Brazil. A rising number of domestic flights has triggered a corresponding rise in the number of aircraft miles flown, along with the number of aircraft takeoffs and landings. Growth in the domestic economy has fueled the construction of additional landing strips, thus allowing more routes to service the population and business centers.[22] Another factor is MROs' new contracts with the Brazilian military, which has led to entire aircraft fleets being overhauled to incorporate up-to-date electronic navigation systems.[23]

Motor Vehicle Parts

The motor vehicle repair industry includes engine remanufacturing and other repair activities. In 2010, the motor vehicle repair industry in Brazil was estimated to be $1.4 billion (R$2.8 billion), concentrated primarily in the state of São Paulo.[24] There are an estimated 2,000 engine remanufacturers in Brazil, half of which are located in São Paulo state. [25] Sixty percent of Brazil's engine remanufactures are reportedly small businesses employing fewer than 6 workers, while the remainder is considered to be SMEs employing between 20 and 60 workers.[26] Conarem, a Brazilian trade association representing engine rebuilders and remanufacturers, represents 350 firms that employ 45,000 workers in Brazil.[27]

Although most of Brazil's engine remanufacturers are small, several large multinational firms account for the majority of remanufacturing activity. For instance, motor vehicle parts producers ZF Sachs Automotive (German ownership), Eaton Corp. (U.S. ownership), and Siemens (German ownership) reportedly account for 75 percent of the total value of Brazil's remanufactured engines and parts. [28] According to industry representatives, many independent motor vehicle engine and parts remanufacturers have not been able to keep pace with the growing demand for aftermarket motor vehicle parts in Brazil because of their small scale (São Paulo state has the largest concentration of motor vehicles, estimated at 21 million).[29] As a result of rising demand for engines and aftermarket parts, OEM producers such as Cummins (U.S. ownership), Mercedes-Benz (German ownership), and MWM/Navistar (U.S. ownership) are reportedly remanufacturing their own engines and parts for the motor vehicle aftermarket.[30]

HDOR Equipment

Brazil's economic growth has bolstered activities in infrastructure development, construction, mining, and energy exploration, all of which are driving demand for HDOR equipment and aftermarket parts (including remanufactured parts) in the country. HDOR equipment producers in Brazil include multinational OEMs such as Caterpillar (U.S. ownership), Cummins (U.S. ownership), Komatsu (Japanese ownership), and Odebrecht (Brazilian ownership). These firms remanufacture diesel engines and core components and provide remanufacturing services to serve OEMs in the HDOR, motor vehicle parts, and industrial machinery sectors. [31] Komatsu is a major producer in Brazil of remanufactured diesel engines and parts (including cylinder heads and piston pumps), as well as construction and industrial machinery.[32]

IT Products

Remanufacturing in the IT products sector in Brazil is primarily limited to printer cartridge remanufacturing. There are an estimated 18,000 firms that manufacture or repair printer cartridges in Brazil. Cartridge remanufacturers reportedly account for about 25 percent of the number of firms within this broader cartridge industry.[33] Most are reportedly small independent third-party firms that remanufacture cartridges produced by leading copier, printer, and cartridge OEM producers in Brazil, which include Katun (U.S. ownership), Lexmark (U.S. ownership), and OKI Data (Japanese ownership).[34]

Brazil reportedly consumes more than 12 million printer cartridges (including remanufactured cartridges) each year, creating a $260 million market.[35] OEM cartridge producers are estimated to account for half of the market, while cartridge remanufacturers account for the remainder. Increasing customer awareness of lower prices for remanufactured cartridges has contributed to rapid growth in the demand for remanufactured printer cartridges in Brazil. However, concerns regarding the quality of remanufactured cartridges have reportedly tempered growth prospects for Brazilian cartridge remanufacturers. [36]

Several of the largest OEMs reportedly try to restrict the supply of used printer cartridges to third-party cartridge remanufacturers by offering discounted prices on OEM replacement cartridges to buyers in return for their used cartridges. Some OEMs contend that third-party cartridge remanufacturers may not restore OEM cartridges to original working condition, and instead sell inexpensive replacement cartridges that have not been fully cleaned or whose components have not been fully replaced, leading to toner leakage and printer damage.[37] Indeed, industry observers note that low quality and a lack of technical know-how have undermined the credibility of independent cartridge remanufacturers in Brazil, but state that quality continues to improve.[38]

Measures Affecting Remanufacturers

The Brazilian government does not distinguish between used and remanufactured goods for import classification purposes. The government imposes a series of restrictions on imports of used capital equipment, which includes both remanufactured goods and cores.[39] As a result of these policies, imports of cores destined to be remanufactured are effectively banned, and imports of remanufactured goods are reportedly negligible owing to the stringent declaration and licensing requirements.

Imports of used capital goods in Brazil are subject to non-automatic licensing. [40] In general, imports of used machinery, equipment, and other capital goods are allowed into Brazil only if they are not produced or otherwise domestically available,[41] and if the age of the goods in question does not exceed the limit of the good's average useful life.[42] In addition to these requirements, imports of remanufactured goods are allowed into Brazil provided that the remanufacturing of the good in question has been performed by the original manufacturer and that the imported good contains the same warranty as an equivalent new product.[43]

Because cores destined to be remanufactured are unlikely to meet the requirements noted above, cores are effectively banned from import into Brazil. As a result, remanufacturers located in Brazil, including Cummins (U.S. ownership), are limited to remanufacturing domestically sourced cores. In certain cases, domestically sourced cores are subject to additional requirements. For instance, engine serial numbers are reportedly linked to vehicle registration, and engines must be de-registered by the vehicle owner before the engine core can be remanufactured.[44] If the remanufacturer is unable to persuade the vehicle owner to complete the deregistration process, engine cores are effectively left in "limbo" until the deregistration process is complete.[45]

In 2002, Brazil approved a law that regulates the import of remanufactured medical devices and equipment.[46] Firms reportedly must comply with a rigid set of guidelines, including providing the date of remanufacturing and a certification of the accurate adjustment and calibration of the equipment. The remanufactured medical equipment must meet the same performance requirements as new medical equipment. Additionally, the OEM must provide technical service assistance in Brazil or designate a local representative to provide the service.[47] In 2011, Brazil announced a proposed regulation that would prohibit the import of remanufactured medical devices altogether, limiting domestic remanufacturing to used medical equipment originally produced in Brazil or imported as new.[48] Leading multinational firms producing medical equipment in Brazil that would likely be adversely affected by the proposed regulation include GE Healthcare (U.S. ownership) and Covidien (Irish ownership).

In 2011, the Brazilian government surveyed the domestic industry to gather information on domestic remanufacturing activities in order to help formulate domestic policies related to the development of Brazil's remanufacturing sector. That research is reportedly ongoing.[49]

The U.S. government continues to engage the government of Brazil on market access for remanufactured goods, particularly through the bilateral U.S.-Brazil Commercial Dialogue's Trade Facilitation Working Group, which includes representatives from the U.S. Department of Commerce (USDOC) and Brazil's Ministry of Development, Industry and Foreign Trade (MDIC). In 2011, the commercial dialogue launched a sustainable supply-chain initiative to share information and best practices and to discuss possible paths to open the Brazilian market.[50] This initiative led to two workshops held in 2011 (in Memphis, TN, and São Paulo, Brazil) to further facilitate an exchange of best practices covering remanufacturing and environmentally friendly product design, and to further promote lowering barriers to trade in remanufactured goods in Brazil.[51]

India

Remanufacturing in India is relatively underdeveloped, with repair activities that restore products to a suitable or working condition being more common. Nevertheless, remanufacturing does occur, primarily in the IT products (printer cartridges) and HDOR equipment sectors. Foreign-based OEMs that have invested in remanufacturing operations in India include Cummins (U.S. ownership) and Volvo (Swedish ownership). India allows the production of remanufactured goods for domestic consumption, but restricts trade in remanufactured goods and related cores, effectively creating two production streams—one for export and the other for domestic consumption. There is no official definition of remanufactured goods for purposes of import declarations, and arbitrary and arcane import licensing requirements make importing remanufactured goods a difficult undertaking.

Remanufacturing Industries and Markets

IT Products

Remanufactured printer cartridges reportedly account for the bulk of remanufactured IT products in India. The sector is largely unregulated, quality varies, and counterfeiting is reportedly common, making the meaning of the term "remanufacture" uncertain. More than 30,000 businesses are reportedly engaged in some form of cartridge refilling or remanufacturing, although only about 70 firms are believed to remanufacture printer cartridges under reputable brands.[52]

The Indian market for printer cartridges is estimated at roughly $250 million annually, with genuine remanufactured printer cartridges estimated to account for less than 10 percent of the total.[53] Given the prevalence of counterfeiting, it is not surprising that negative customer perceptions of remanufactured printer cartridges are reportedly the sector's biggest challenge. However, efforts by remanufacturers to increase customer awareness of the environmental benefits of remanufactured goods have reportedly helped improve customers' attitudes.[54] The Indian Cartridge Remanufacturers and Recyclers Association (ICRRA) and the Cartridge Recyclers and Traders Association of India (CRTAI) are trade associations formed with the assistance of the Indian government to promote domestic production of remanufactured printer cartridges for both the Indian market and for export. ICRRA and CRTAI's goals are to promote the use of remanufactured printer cartridges in India, strengthen the role of the industry in the nation's economic development, and promote product quality.[55]

HDOR Equipment

Infrastructure development and construction activity, fed by economic growth, are principal drivers of demand for HDOR equipment in India. Demand for HDOR equipment has in turn bolstered demand for engines and aftermarket parts, including remanufactured HDOR equipment.[56] Volvo recently began remanufacturing construction equipment at its facility in Bangalore for the domestic market.[57] Cummins operates two separate remanufacturing facilities—one for exports and the other for the domestic market—to comply with Indian laws that restrict the production of remanufactured goods for the domestic market made from imported cores, as described below. [58] (At the Commission hearing, a company representative for Cummins noted that running both facilities has resulted in operating inefficiencies and higher costs.)[59]

Measures Affecting Remanufacturers

India generally does not distinguish between used and remanufactured goods for import classification purposes. The country prohibits imports of used goods (cores) to be remanufactured in India and sold in the domestic market, but allows imports of cores that are intended to be remanufactured domestically and subsequently exported. Barring a few exceptions, imports of used capital goods (including refurbished or reconditioned spare parts or components) intended for the domestic market are prohibited unless it can be shown that such goods retain 80 percent of the residual value of the original (new) products (pending certification by a chartered engineer).[60] Because cores destined to be remanufactured are unlikely to retain this level of value, cores intended to be remanufactured and sold domestically are effectively banned from import into India. Imports of used computers and laptops are banned outright.

Imports of remanufactured goods are subject to licensing, even though India's foreign trade policy document contains no official definition of remanufactured goods. [61] Determining what constitutes a remanufactured good versus a used one is thus open to interpretation, and there are few guidelines or objective criteria to allow a license issuer (or an importer) to distinguish between a remanufactured, refurbished, or reconditioned good. In addition, licensing requirements are reportedly onerous and border on the arcane. For instance, the import license reportedly allows manufacturers in India to import only the specific remanufactured parts listed on the license. The parts-specific licensing requirements, coupled with long approval times to obtain a new license, can hamper the ability of importers to either import more or change the product mix of import shipments in response to evolving demand. Reportedly, there has only been one such license issued to Caterpillar, a producer of both new and remanufactured HDOR equipment. A company representative at the Commission hearing noted that Caterpillar was previously granted a license to import remanufactured goods into India, but that the Indian government recently rejected a new license application, citing concerns over product quality. Caterpillar noted that their Indian distributors offer warranties on all sales of remanufactured goods. Nevertheless, the license was still rejected.[62]

China

China's remanufacturing industry is of relatively recent origin, only having been formally established in 2008. Remanufacturing activities are primarily regulated through two government-led pilot programs—one for the motor vehicle parts remanufacturing sector and another for the industrial machinery and electrical equipment remanufacturing sector. Only firms that are approved to participate in these programs may remanufacture. China maintains restrictive measures on foreign trade in remanufactured goods and related cores. Statistics on Chinese remanufacturing production, foreign trade, and employment are largely unavailable, due in part to the relatively young age of the industry.

Overview of Government Policies

The development of China's remanufacturing industry is an outgrowth of national policies and laws aimed at reducing environmental pollution and promoting recycling and sustainability. In 2001, the Chinese government created the National Key Laboratory for Remanufacturing Technology to focus on developing remanufacturing technologies for both commercial and military equipment.[63] Beginning in 2003, a series of laws were enacted to address pollution, recycling, and energy conservation,[64] ending with the latest law (2009), which expanded the application of remanufacturing.[65] Beginning in 2008, China's National Development and Reform Commission (NDRC) and the Ministry of Industry and Information Technology (MIIT) established two pilot programs (described below) which allow limited remanufacturing in the motor vehicle parts sector, and in the industrial machinery and electrical equipment sector.[66]

In 2010, the NDRC and 11 other ministries and agencies issued a guidance document for promoting the development of China's remanufacturing industry.[67] The document listed the industries and particular products that should be promoted for remanufacturing, established broad goals for remanufacturing, noted the major challenges, and presented a strategic road map for implementation.[68] Additional guidance released in September 2011 sought to ensure that remanufacturing pilot projects achieved tangible results and to expand the number of industries approved for remanufacturing.[69] Despite broad central government support, the lack of standard terminology has reportedly limited progress among ministries in developing rules and regulations governing China's remanufacturing sectors.[70]

Policies Affecting the Development of the Motor Vehicle Parts Remanufacturing Sector

China's initiatives for remanufacturing motor vehicle parts are influenced by the strategic nature and importance of the motor vehicle industry and the desire to promote industrial and environment sustainability. The pilot program, launched in 2008, authorizes the remanufacture of engines, transmissions, generators, starters, drive shafts, compressors, oil pumps, water pumps, and other components.[71] Approximately 15 motor vehicle parts remanufacturers are approved to take part in the pilot program. Foreign involvement is somewhat limited. Cummins (U.S. ownership), Kia (Korean ownership), and Volkswagen (German ownership) have all reportedly established remanufacturing joint ventures with their respective Chinese partners. According to the China Association of Automobile Manufacturers, in 2009 China remanufactured approximately 110,000 engines, 60,000 transmissions, and 1 million starter motors.[72]

However, contradictory government policies and overlapping authorities from different government entities have reportedly hampered the development of the motor vehicle parts remanufacturing sector in China at times. For instance, Decree No. 307 of the State Council, enacted in August 2001, required that five motor vehicle part assemblies (engine, power transmission box, steering, axle, and chassis) be destroyed and recycled as scrap metals at the end of a vehicle's life instead of being recovered for reuse. [73] However, in the announcement creating the motor vehicle parts remanufacturing pilot program, the NDRC stated that the program would allow for the remanufacture of engines, transmissions, and steering assemblies. Nevertheless, firms in the pilot program were temporarily prohibited from purchasing these parts from dismantlers of end-of-life motor vehicles.[74]

Policies Affecting the Industrial Machinery and Electrical Equipment Remanufacturing Sector

In 2009, the MIIT established the industrial machinery and electrical equipment remanufacturing pilot program, which allows the controlled remanufacture of construction machinery; industrial, electrical, and mechanical equipment; machine tools; mining machinery; railway locomotives and equipment; marine equipment; office equipment; and IT products. Approximately 60 remanufacturers were approved to take part in the program.[75]

Foreign investment in industrial machinery and electrical equipment remanufacturing has been somewhat limited. Only five foreign firms (two from the United States, two from Japan, and one from Korea) reportedly remanufacture industrial machinery and electrical equipment. Caterpillar (U.S. ownership) formed a joint venture with China Yuchai International Ltd. through China Yuchai International's principal subsidiary, Guangxi Yuchai Machinery Co. Ltd., establishing Yuchai Remanufacturing Services (Suzhou) Ltd. to provide remanufacturing services for Yuchai diesel engines and components and certain Caterpillar diesel engines and components for customers worldwide.[76] Caterpillar also has its own operations in which it remanufactures pumps and motors (hydraulic pumps, fuel pumps, oil pumps, and water pumps) and engine components (cylinder packs, cylinder heads, and injector nozzles), and provides remanufacturing services to other Caterpillar business units and other external companies in China.[77] In 2010, Fuji Xerox (Japanese ownership) received certification to operate a pilot remanufacturing operation at its printer parts and cartridge recycling facility in Suzhou.[78]In February 2012, IBM (U.S. ownership) began remanufacturing computer network servers in Shenzhen, being the first IT company to do so in China.[79]

Policies Affecting Trade in Remanufactured Goods in China

China maintains measures that either prohibit or restrict foreign trade in remanufactured goods and related cores for remanufacturing. Regulations and restrictions vary by product. However, there is no agreed definition or standard for remanufactured goods in China, and remanufactured goods are typically classified as "old" or "scrap" products. In theory, foreign remanufactured products may be imported into China on the condition that reassembly occurs there. However, for practical purposes, importation of remanufactured products is considered virtually impossible. In addition, China's customs service does not maintain a separate classification for remanufactured goods, making it difficult to apply the correct import duty.[80]

Policies Affecting Trade in Remanufactured Motor Vehicle Parts

In the motor vehicle parts sector, China allows the importation of certain used motor vehicle parts (cores), including engines and transmissions, into export processing zones (EPZs) for remanufacturing and subsequent export. However, China prohibits the importation of cores directly into China's customs territory to be remanufactured domestically.[81] As a result, motor vehicle parts (and HDOR equipment) remanufacturers that sell directly into the Chinese market are dependent on domestic core supplies.

At the Commission hearing, a Caterpillar representative stated that the firm would increase its production of remanufactured goods in China if it could import cores to supplement those obtained from the Chinese market.[82] Both Caterpillar and Cummins still consider that significant remanufacturing opportunities exist in China because of the large size of the Chinese engine and machinery markets.[83]

Policies Affecting Trade in Remanufactured Industrial Machinery and Electrical Equipment

China's Ministry of Commerce (MOFCOM) maintains specific policies on imports of mechanical and electronic products, including used products. Remanufactured mechanical and electronic products are considered "used" products for import purposes, and are classified as either prohibited, restricted subject to licensing, subject to automatic licensing, or imported freely, depending upon the product. Used products classified in the restricted category, which may include remanufactured goods, require an import license. Another set of regulations governing imports of used mechanical and electronic products to be remanufactured stipulates that the importer/end user must be licensed and approved to perform remanufacturing in the industrial machinery and electrical equipment pilot program established by MIIT.[84]

Policies Affecting Trade in Other Remanufactured Products

Regulations governing the importation of solid wastes likely restrict imports of other products used as cores for remanufacturing. For example, in 2008 China's Ministry of Environmental Protection (MEP) issued regulations that prohibit the importation into both China's customs territory and EPZs of goods that have lost their original value, have been discarded, or are otherwise considered solid waste. The catalog of prohibited imports includes tire casings and certain used equipment and parts.[85] Prohibited items likely include certain heavy-duty engines; transmissions and other parts of construction and agricultural machinery; electric motors; medical equipment; and other products that might be used as cores for remanufacturing.

China reportedly bans imports of remanufactured medical devices because they are classified as used medical devices, which are prohibited.[86] At the Commission hearing, GE Healthcare stated that if China did not have such a ban, the firm would likely import remanufactured medical devices into China.[87]

U.S.-China Remanufacturing Dialogue

The United States and China are engaged in an ongoing dialogue to address China's restrictive policies regarding remanufacturing. The two countries held a Remanufacturing Dialogue in October 2011, and remanufacturing was also discussed at the United States-China Joint Commission on Commerce and Trade meeting held in November 2011.[88]

Other Remanufacturing Sectors in China

Retreaded Tires

Tire retreading is an established industry in China, although it appears to be a somewhat unregulated activity that falls outside the scope of China's remanufacturing pilot programs. Nevertheless, there are reportedly over 1,000 firms that retread tires in China, although retreading activity is concentrated among 55 leading tire retreaders.

Approximately 50 firms in China each produce more than 50,000 retreaded tires annually, although only a handful of firms produce more than 100,000 retreaded tires annually, and only a single firm reportedly produces more than 300,000 retreaded tires per year.[89] Total Chinese production of retreaded tires was estimated at 11.5 million tires in 2008, and was forecast to reach 15 million tires in 2009 and 19 million tires in 2010.[90] In comparison, approximately 15.5 million retreaded truck tires were produced in the United States in 2011, as noted in chapter 8.

Some smaller firms focus solely on retreading activities, and are reportedly hampered by a limited supply of suitable tire casings.[91] Factors limiting supply include poor tire quality, the lack of an adequate scrap tire collection system,[92] and a ban on imports of tire casings. Weak enforcement of technical quality standards for new tires and a lack of inspections to catch tire wear beyond maximum limits also reportedly play a role.[93]

In contrast, larger integrated tire producers, including both Chinese domestic producers and, to a lesser extent, foreign subsidiaries of global multinational tire producers, have entered the tire retreading industry and benefited from a captive supply of tire casings under their own banner. Foreign investment is rather limited—Michelin (French ownership) opened a tire retreading plant in September 2005 and Bridgestone (Japanese ownership) opened an aircraft tire retread plant for production and sale of aircraft tires in July 2007.[94] Demand in China for OTR retreaded tires is expected to increase in response to economic growth and increased construction activity. Demand for retreaded tires from urban transit vehicles, such as bus fleets, is also reportedly strong.[95]

Other Foreign Countries

Remanufacturing operations located elsewhere in the world are relatively limited in size and scope and tend to be co-located with existing manufacturing operations. In other countries such as Singapore and Korea, described below, remanufacturing has been recognized by the government within the context of broader efforts to promote sustainability and reduce waste in industrial activities.

Singapore

Remanufacturing in Singapore occurs in the HDOR equipment, motor vehicle parts, medical devices, electrical apparatus, and marine equipment remanufacturing sectors. The Singapore government does not distinguish between remanufactured and other goods (new or used) and does not impose specific marking or labeling requirements for remanufactured goods. Cores to be remanufactured are imported freely, and remanufactured products are sold domestically and exported regionally to Association of Southeast Asian Nations (ASEAN) countries and Australia. [96] In 2011, Caterpillar (U.S. ownership) opened a facility in Singapore to remanufacture equipment for OTR trucks and mining work, including transmissions, drives, and torque converters, to serve the regional mining industry.[97] Eighty percent of Caterpillar's products remanufactured in Singapore are reportedly exported to Indonesia, with the rest exported to Australia. Another firm that recently established remanufacturing operations in Singapore is Tru-Marine, a remanufacturer of turbocharger components used in marine, power plant, and locomotive applications.[98] Although Singapore does not collect specific data on domestic remanufacturing production or trade in remanufactured goods, the government is committed to establishing a remanufacturing base and related R&D in an effort to increase value-added activities in the country's manufacturing sector. In 2011, the government launched the Advanced Remanufacturing and Technology Center, an R&D center that works with local universities and remanufacturers to develop remanufacturing technologies for the aerospace, motor vehicle parts, marine, and HDOR equipment sectors.[99] The center has partnered with a handful of SMEs and larger multinational companies, including Boeing (U.S. ownership), Rolls-Royce (UK ownership), and Siemens (German ownership), among others.

Korea[100]

Remanufacturing in Korea is reportedly limited, but occurs in the motor vehicle parts, HDOR equipment, IT products (primarily printer cartridges), medical devices, and military defense sectors. The motor vehicle parts sector reportedly accounts for the vast majority (95 percent) of domestic remanufacturing activities. The industry is relatively fragmented, with 1,500–2,000 SMEs remanufacturing motor vehicle parts. Annual production of remanufactured vehicle parts is estimated to be about $217 million. About 200–250 companies, mostly small ones, remanufacture printer toner cartridges, with an estimated annual production of $71 million. The market for remanufactured HDOR equipment is reportedly small, due in part to Korean construction firms' contracting out repair activities on their HDOR equipment to independent suppliers, instead of relying on OEMs to remanufacture OEM-branded equipment. Excavators make up 40 percent of remanufactured HDOR equipment in Korea. In 2011, the Korean government identified remanufacturing as an industrial activity that could foster sustainable growth in "green" industries, create employment, and stabilize prices in the country. The Korean Institute for Industrial Economics estimates that remanufacturing can conserve 85–90 percent of the materials and energy used to manufacture new products.

Other

In other APEC countries, such as Canada and New Zealand, remanufacturing is limited to a few industry sectors, and there are no government policies that either encourage or discourage remanufacturing activities and trade. Table 10.1 provides an overview of remanufacturing sectors and measures affecting remanufacturers in selected APEC countries that have not yet been discussed and for which information is available.

Table 10.1. Remanufacturing sectors and measures affecting remanufacturers in selected APEC countries

Country	Remanufacturing sectors	Measures affecting remanufacturers
Canada	Motor vehicle parts, aerospace components, medical devices and equipment.	The government does not distinguish between new and remanufactured goods; there are no official government statistics on trade in remanufactured goods.
Malaysia	Motor vehicle parts (estimated to be $7 million annually), MRO activities.	The government prohibits imports of used motor vehicle parts for reuse. Imported cores to be remanufactured domestically are thus likely prohibited as well.
Mexico	Consumer electronics, IT products (printers and printer cartridges), and motor vehicle parts. There are approximately 60 known remanufacturers in Mexico.	There are no specific laws that regulate remanufacturing as an industrial activity. However, in 2008, Mexico implemented the "Three Rs" (reduce, reuse, and recycle) initiative to promote sustainability and to reduce waste in industrial activities.
New Zealand	Motor vehicle parts, machinery, IT products (printer cartridges).	High labor costs are reportedly a deterrent to establishing remanufacturing operations in New Zealand.
Russia	HDOR equipment, IT products (copier machines and printer cartridges).	There is no direct legal restriction on remanufactured products, but complicated customs processes and paperwork requirements make core returns to parts distribution centers expensive.
Vietnam	Apparently very little remanufacturing activity in any sector.	Regulations ban the import or export of second-hand goods, of which remanufactured goods have been expressly defined as a subset.

Sources: Compiled by USITC staff from U.S. embassy cables; USITC, hearing transcript, February 28, 2012, 165 (testimony of Pierce, Cummins Inc.); Cummins prehearing submission to the USITC, February 9, 2012.

REFERENCES

Boustani, Melaz, Sartaj Sahni, Stephen Graves, and Tim Gutowski. "Appliance Remanufacturing and Life Cycle Energy and Economic Savings." *Sustainable Systems and Technology,* Spring 2010.

Bridgestone Corp. "Bridgestone Officially Opens Aircraft Tire Retreading Plant in China." New release, July 25, 2007.

Caterpillar, Inc. "Caterpillar Announces Remanufacturing Joint Venture with China Yuchai to Promote China's Sustainability and Environmental Preservation Initiatives." Press release, December 14, 2009.

———. "Caterpillar in China: Shanghai," n.d. http://china.cat.com/cda /components/fullArticle?m=315915&x=7&id=2778000 (accessed December 2, 2011).

Chapman, Adrian, Caroline Bartlett, Ian McGill, David Parker, and Ben Walsh. *Remanufacturing in the UK: A Snapshot of the UK Remanufacturing Industry, 2009.* Resource Recovery Forum, 2010. http://www.remanufacturing.org.uk/pdf/story /1p342.pdf.

Chen, Ming and Fan Zhang. "End-of-Life Vehicle Recovery in China: Consideration and Innovation following the EU ELV Directive." JOM 61, no. 3 (March 2009).

China National Heavy Duty Truck Group Jinan Fuqiang Power Co., Ltd., "About Us," n.d. http://fuqiangpower.en.alibaba.com/aboutus.html (accessed December 22, 2011).

Cummins, Inc. Written submission to the U.S. International Trade Commission in connection with inv. no. 332-525, *Remanufactured Goods: An Overview of the U.S. and Global Industries, Markets, and Trade,* February 9, 2012.

Datamonitor. "Caterpillar Inc.: Company Profile," September 20, 2011.

———. "Komatsu Ltd: Company Profile," June 2, 2011.

———. "Lexmark International, Inc.: Company Profile," August 17, 2011.

Ecostar. "State General Administration of AQSIQ Visited ECOSTAR." Press release, October 24, 2005. http://en.ecostar.com.cn/About/News/news_ show.asp?cid=58.

Fuji Xerox, Co., Ltd. "Chinese Government Certifies Fuji Xerox Recycling Site as a Pilot Remanufacturing Enterprise of Electric Products Fuji Xerox," February 2, 2010.

Geetha, Nandikotkur. "Remanufacturing Growth," *Channel Business*, n.d. http://www. channelbusiness.in/index.php?option=com_content&task=view&id=355&Itemid=78 (accessed September 2, 2012).

Government of China. Ministry of Commerce (MOFCOM). "Measures for Administration of Import of Specified Used Mechanical and Electronic Products." No. 5 [2008], April 7, 2008. http://english.mofcom.gov.cn /aarticle/policyrelease/announcement/200805/20080 505551892.htm l.

———. Ministry of Commerce (MOFCOM). "Catalog for 2012 of Goods Subject to Import Licenses Issued by Designated Administrations." Notice No. 102 [2011], December 31, 2011. http://www.mofcom.gov.cn/aarticle /b/e/201201/20120107914675.html.

———. Ministry of Commerce (MOFCOM), General Administration of Customs (GACC), General Administration of Quality Supervision, Inspection and Quarantine (AQSIQ). "Import License Management Catalog of Imports 2012." Notice No. 99 [2011], December 30, 2011. http://www.mofcom.gov.cn/aarticle/b/e/201112/20111 207910199. html.

———. Ministry of Industry and Information Technology (MIIT) and National Development and Reform Commission (NDRC). *Automotive Industry Development Policy*, September 1, 2009.

———. Ministry of Industry and Information Technology (MIIT). "Ministry of Industry and Information Technology Issues Notices, 'Pilot Remanufacturing of Electromechanical Products and First List of Enterprises' and 'Mechanical and Electrical Products Pilot Program Requirements'." No. 663 [2011], December 21, 2009. http://www.miit. gov.cn/n11293472/n11293832/n12843926/12931388.html (accessed October 26, 2011).

———. National Development and Reform Commission (NDRC). "Guidelines for Promoting the Development of the Remanufacturing Industry." No. 991 [2010], May 31, 2010. http://www.ndrc.gov.cn /zcfb/zcfbtz/2010tz/t20100531 350189.htm.

———. National Development and Reform Commission (NDRC). "Notice on Experimental Work on Deepening Remanufacturing." No. 2170 [2011], September 14, 2011. http://www.ndrc.gov.cn/zcfb/zcfbtz/2011tz /t20110914_433941.htm.

———. State Council. "Decree No. 307 of the State Council of the People's Republic of China: Measures for Administration of Recycling of Scrapped Motor Vehicles." *Gazette of the State Council of the People's Republic of China,* no. 23, serial no. 1022, August 20, 2001. http://www.gov.cn /gongbao/content/2001/content 60919.htm.

Government of India. Ministry of Commerce and Industry, Department of Commerce. *Foreign Trade FTP, 2009–2014*, August 23, 2010. http://dipp. gov.in/English/Policies/ National Manufacturing FTP 25October 2011.pdf.

———. Ministry of Commerce and Industry, Department of Commerce. *Hand Book of Procedures (Vol. I)*, 2009–2014, effective August 27, 2009.

Gray, Casper and Martin Charter. "Remanufacturing and Product Design: Designing for the 7th

Generation." Farnham, UK: The Center for Sustainable Design, University College for the Creative Arts, 2007.

Grose, Thomas K. "Born Again." *Time International (South Pacific Edition)* 28, July 12, 2007. http://www.time.com/time/magazine /article /0,9171,1642728,00.html.

Holeung Ho Lee Foundation. "Xu Binshi," 2006. http://www.hlhl.org.cn /english x/showsub.asp?id=733. IBM. "IBM Opens the First Server Remanufacturing Center in China." Press release, February 29, 2012.

Lemmel, M.C. Written submission from the International Federation of Engine Rebuilders and Remanufacturers to the European Commission's Internet Consultation within the Framework of the CARS21 Activities, April 2005. http://ec.europa.eu/enterprise/sectors /automotive/documents/consultations/2005-next10/index en.htm.

M2Presswire. "TAM Airlines Registers an Operating Profit of R$977 Million in 2010," March 10, 2011.

Morrison, Murdo. "How Portugal's TAP M&E Became Brazil's Biggest MRO." *Flight Global*, May 22, 2012. http://www.flightglobal.com/news/articles/in-focus-how-portugals-tap-mebecame-brazils-biggest-mro-371720/.

NBM Media. "Volvo Outlines Growth Plans for India," May 2011.

Oliveria, Flavio. "The State of the Remanufacturing Industry in Brazil." *Recharger Magazine,* October 11, 2011.

Parker, David. Remanufacturing in the UK: A Significant Contributor to Sustainable Development? UK: Oakdene Hollins Ltd., 2004.

———. An Analysis of the Spectrum of Re-use: A Component of the Remanufacturing Pilot for Defra, BREW Programme. UK: Oakdene Hollins Ltd., May 2007.

Pierce, Allen. Cummins, Inc. Written submission to the U.S. International Trade Commission in connection with inv. no 332-525, *Remanufactured Goods: An Overview of the U.S. and Global Industries, Markets, and Trade*, February 9, 2012.

PR Newswire. "Caterpillar Announces Opening of New Singapore Remanufacturing Facility," May 24, 2011.

PR Web. "Singapore Launches Asia's Inaugural Remanufacturing Industry in Singapore," June 22, 2012. *Recharger Magazine*. "Reciclamais South American Expo Promotes Brazilian Market," February 1, 2008.

Retreading Business. "Chinese Retread Volume to Increase by 30%," issue 2009/3. http://www.retreadingbusiness.com/features/chineseretread.asp.

———. "Interview: China Tyre Retreading, Repairing and Recycling Association," issue 2009/2. http://www.retreadingbusiness.com/features /ctra.asp.

———. "Zhu Jun Explains the Current Status of the CTRA," issue 2010/1. http://www.retreadingbusiness.com/features/zhujun.asp.

Sina.com.cn. "Remanufacturing: Looks Beautiful," March 31, 2012. http://finance.sina.com.cn/leadership/msypl/20110321/18209568127.shtml.

Sobie, Brendan. "Brazil MRO Sector Poised for Major Expansion." *Flight International,* July 6, 2010. http://www.flightglobal.com/news/articles /brazil-mro-sector-poised-for-major-expansion-343930.

ThisIsGloucestershire. "Xerox Is Now a Giant PO Box for UK Firms," March 11, 2010. http://www.thisisgloucestershire.co.uk/Xerox-giant-PO-box-UK-firms/story-11858201-detail/story.html.

U.S. Department of Commerce (USDOC). International Trade Administration (ITA). *Global Import Regulations for Pre-Owned (Used and Refurbished) Medical Devices.* 6th ed., May 2008. http://www.ita.doc.gov/td/health/Pre OwnedMedEquipment FINAL 050608.pdf.

U.S. Department of State. U.S. Embassy, Beijing, "China's Remanufacturing Market (BEIJING 000460)," February 1, 2012.

———. U.S. Embassy, Brasilia. "Brazil: Response to ITC Inquiry on Remanufactured Goods (BRASILIA 000232)," February 12, 2012.

———. U.S. Embassy, London. "United Kingdom: Remanufactured Goods Market—USITC Request (LONDON 000344)," January 30, 2012.

———. U.S. Embassy, Madrid. "Madrid Response to USITC Conducting Fact-Finding Investigation of the U.S. and Global Industries, Markets, and Trade for Remanufactured Goods (MADRID 000072)," January 25, 2012.

———. U.S. Embassy, Paris. "Embassy Paris Response to USITC Request for Information on Remanufactured Goods (PARIS 000066)," January 18, 2012.

———. U.S. Embassy, Seoul. "ROK Remanufacturing Market (SEOUL 000114)," March 9, 2012.

———. U.S. Embassy, Singapore. "An Overview of the Remanufacturing Industry in Singapore (SINGAPORE 000058)," March 9, 2012.

———. U.S. Embassy, Sofia. "Remanufactured Medical Equipment, Automotive Parts, and Airplanes (SOFIA 000010)," January 12, 2012.

———. U.S. Embassy, The Hague. "The Netherlands: Remanufactured Goods Market Very Limited (THE HAGUE 000046)," January 12, 2012.

U.S. International Trade Commission (USITC). Hearing transcript in connection with inv. no. 332-525, *Remanufactured Goods: An Overview of the U.S. and Global Industries, Markets, and Trade,* February 28, 2012.

United States Trade Representative (USTR). "2011 U.S.-China Joint Commission on Commerce and Trade Outcomes." Fact sheet, November 21, 2011. http://www.ustr.gov/

about-us/press-office/fact-sheets/2011/november/2011-us-china-joint-commission-commerce-and-trade-ou.

———. "Import Licensing," n.d. http://www.ustr.gov/trade-agreements/wto-multilateral-affairs/wtoissues/import-licensing (accessed September 3, 2012).

———. "Outlines of the Trans-Pacific Partnership Agreement," November, 2011. http://www.ustr.gov/about-us/press-office/fact-sheets/2011 /novem ber/outlines-trans-pacific-partnership-agreement.

Xiang, Wang, and Chen Ming. "Implementing Extended Producer Responsibility: Vehicle Remanufacturing in China." *Journal of Cleaner Production* 19 (2011), 680–86.

APPENDIX A: REQUEST LETTER

EXECUTIVE OFFICE OF THE PRESIDENT
THE UNITED STATES TRADE REPRESENTATIVE
WASHINGTON, D.C. 20508

DOCKET NUMBER 2825
Office of the Secretary
Int'l Trade Commission

JUN 2 8 2011

The Honorable Deanna Tanner Okun
Chairman
U.S. International Trade Commission
500 E Street, SW
Washington, DC 20436

Dear Chairman Okun:

Remanufacturing is an industrial process that restores end-of-life goods to original working condition and occurs across a range of sectors, such as heavy machinery, automotive parts, medical equipment, office equipment, information and communications technology equipment, and precision instruments. Remanufacturers include original equipment manufacturers, independent companies, and small and medium size enterprises (SMEs). American consumers and businesses purchase remanufactured products and use them in their daily activities, as they know that such products provide high-end performance at lower cost. Remanufacturing also provides significant environmental benefits, since it uses less material, energy, and water than manufacturing goods from virgin material.

U.S. remanufacturers have been leaders in remanufacturing for decades and are expanding their activities across the globe. Despite the strong presence of remanufacturing in the United States and growth internationally, little data regarding remanufacturing, including sales, employment, trade, and investment in remanufacturing have been compiled and analyzed. I believe that additional data collection and analysis will help remanufacturers, and in particular SMEs, in growing their companies, accessing foreign markets, and responding to increasing resource scarcity. Such information and analysis will also help the United States pursue discussions and negotiations on eliminating barriers to trade in remanufactured goods bilaterally and multilaterally at the World Trade Organization and at Asia-Pacific Economic Cooperation (APEC). As the Administration strives to remove barriers and constraints to U.S. exports, enforce trade agreements, and support the expansion of trade, remanufacturing and the export of remanufactured goods can contribute positively to trade and the U.S. economy.

Pursuant to authority delegated by the President to the United States Trade Representative (USTR) under Section 332(g) of the Tariff Act of 1930, I request that the Commission prepare a report that: 1) provides an overview of the U.S. remanufactured goods industries and markets; 2) estimates U.S. and global trade in remanufactured goods to the extent possible; and 3) examines factors affecting trends in remanufactured goods trade. As the process of remanufacturing may be employed by any industrial sector throughout the U.S. economy, this report should focus on remanufacturing-intensive sectors that account for the majority of remanufacturing activity in the United States. For purposes of its analysis, the Commission should define remanufactured goods as non-agricultural goods that are entirely or partially comprised of parts that (i) have been obtained from the disassembly of used goods; and (ii) have been processed, cleaned, inspected, and tested to the extent necessary to ensure they have been restored to original working condition or better; and for which the remanufacturer has issued a warranty.

Specifically, based on a review of available data and other information, including the collection of primary data through a survey of enterprises engaged in remanufacturing, the Commission's report should cover the period 2009–11 and, to the extent practicable, estimate and describe the following:

- the size and scope of remanufacturing in the United States, including principal producers, levels of employment, investment, and sales (including in the domestic market and exports);
- the U.S. market for remanufactured goods, including the goods supplied by domestic producers;
- U.S. exports of remanufactured goods, including by primary sector and leading export destinations. The report should also compare U.S. exports of remanufactured goods to free trade agreement partners with exports to non-free trade agreement partners. To the extent possible, information should also be included on the level of U.S. imports of remanufactured goods as well as the level of imports of "cores" (used goods acting as the primary component input(s)) and the principal sources for these products; and
- the factors affecting sales, trade, and investment in U.S. remanufactured goods, including a discussion of recent trends.

In addition, USTR would benefit from information regarding remanufacturing in foreign markets (e.g., such as markets in the APEC and G-20 economies), and the role of foreign direct investment (FDI) in the United States for remanufactured goods. Therefore, to the extent possible, the report should also include the following:

- an assessment of FDI in U.S. remanufacturing, including associated employment and the extent to which foreign firms have invested in the U.S. enterprises producing remanufactured goods;
- an overview of outward FDI in remanufacturing by leading U.S.-based firms, including their major markets;
- an analysis of trade by foreign remanufacturers invested in the United States, including trade in "cores"; and
- an overview of the global markets for remanufactured goods and enterprises engaged in remanufacturing.

I anticipate that USTR will make the Commission's report available to the public in its entirety. Therefore, the report should not contain any confidential business or national security information. The Commission is requested to deliver the report no later than 16 months from the date of receipt of this letter.

The Commission's assistance in this matter is greatly appreciated.

Sincerely,

Ambassador Ronald Kirk

Appendix B: Federal Register Notice

INTERNATIONAL TRADE COMMISSION

[Investigation No. 332–525]

Remanufactured Goods: An Overview of the U.S. and Global Industries, Markets, and Trade; Institution of Investigation and Scheduling of Public Hearing

AGENCY: United States International Trade Commission.

ACTION: Notice.

SUMMARY: Following receipt of a request dated and received June 28, 2011 from the U.S. Trade Representative (USTR) under section 332(g) of the Tariff Act of 1930 (19 U.S.C. 1332(g)), the U.S. International Trade Commission (Commission) instituted investigation No. 332–525, Remanufactured Goods: An Overview of the U.S. and Global Industries, Markets, and Trade.

DATES:

January 31, 2012: Deadline for filing requests to appear at the public hearing.

February 9, 2012: Deadline for filing pre-hearing briefs and statements.

February 28, 2012: Public hearing.

March 8, 2012: Deadline for filing post-hearing briefs.

May 20, 2012: Deadline for filing all other written submissions.

October 28, 2012: Transmittal of Commission report to the USTR.

ADDRESSES: All Commission offices, including the Commission's hearing rooms, are located in the United States International Trade Commission Building, 500 E Street, SW., Washington, DC. All written submissions should be addressed to the Secretary, United States International Trade Commission, 500 E Street, SW., Washington, DC 20436. The public record for this investigation may be viewed on the Commission's electronic docket (EDIS) at *https://edis.usitc.gov/ edis3-internal/app.*

FOR FURTHER INFORMATION CONTACT:

Project Leader Alan Treat (202–205–3426 or *alan.treat@usitc.gov*) or Deputy Project Leader Jeremy Wise (202–205– 3190 or *jeremy.wise@usitc.gov*) for information specific to this investigation. For information on the legal aspects of these investigations, contact William Gearhart of the Commission's Office of the General Counsel (202–205–3091 or *william.gearhart@usitc.gov*). The media should contact Margaret O'Laughlin, Office of External Relations (202–205– 1819 or *margaret.olaughlin@usitc.gov*). Hearing-impaired individuals may obtain information on this matter by contacting the Commission's TDD terminal at 202–205–1810. General information concerning the Commission may also be obtained by accessing its Internet server (*http://www.usitc.gov*). Persons with mobility impairments who will need special assistance in gaining access to the Commission should contact the Office of the Secretary at 202–205–2000.

Background: As requested by USTR, the Commission will conduct an investigation and prepare a report that provides an overview of the U.S. remanufactured goods industries and markets, estimates U.S. and global trade in remanufactured goods to the extent possible, and examines factors affecting trends in remanufactured goods trade. The Commission's report will focus on remanufacturing-intensive sectors in the U.S. economy that account for the majority of remanufacturing activity in the United States. For the purpose of its analysis, the

Commission will define remanufactured goods as non- agricultural goods that are entirely or partially comprised of parts that (i) Have been obtained from the disassembly of used goods; and (ii) have been processed, cleaned, inspected, and tested to the extent necessary to ensure they have been restored to original working condition or better; and for which the remanufacturer has issued a warranty. The Commission will base its report on a review of available data and other information, including the collection of primary data through a survey of enterprises engaged in remanufacturing. The report will cover the period 2009–11, and to the extent practicable will estimate and describe the following:

- The size and scope of remanufacturing in the United States, including principal producers, levels of employment, investment, and sales (including in the domestic market and exports);
- The U.S. market for remanufactured goods, including the goods supplied by domestic producers;
- U.S. exports of remanufactured goods, including by primary export sector and leading export destinations. The report will also compare U.S. exports of remanufactured goods to free trade agreement (FTA) partners with exports to non-FTA partners. To the extent possible, the report will also include information on the level of U.S. imports of remanufactured goods as well as the level of imports of "cores" (used goods acting as the primary component input(s)) and the principal sources for these products; and
- The factors affecting sales, trade, and investment in U.S. remanufactured goods, including a discussion of recent trends.

To the extent possible, the report will also include the following:

- An assessment of foreign direct investment (FDI) in U.S. remanufacturing, including associated employment and the extent to which foreign firms have invested in U.S. enterprises producing remanufactured goods;
- An overview of outward FDI in remanufacturing by leading U.S.-based firms, including their major markets;
- An analysis of trade by foreign remanufacturers invested in the United States, include trade in "cores"; and
- An overview of the global markets for remanufactured goods and enterprises engaged in remanufacturing.

The USTR asked that the Commission deliver its report no later than October 28, 2012.

Public Hearing: A public hearing in connection with this investigation will be held at the U.S. International Trade Commission Building, 500 E Street, SW., Washington, DC, beginning at 9:30 a.m. on Tuesday, February 28, 2012. Requests to appear at the public hearing should be filed with the Secretary, no later than 5:15 p.m., January 31, 2012, in accordance with the requirements in the "Submissions" section below. All pre-hearing briefs and statements should be filed not later than 5:15 p.m., February 9, 2012; and all post-hearing briefs and statements should be filed not later than 5:15 p.m., March 8, 2012. In the event that, as of the close of business on January 31, 2012, no witnesses are scheduled to appear at the hearing, the hearing will be canceled. Any person interested in attending the hearing as an observer or nonparticipant should contact the Office of the Secretary at 202–205–2000 after January 31, 2012, for information concerning whether the hearing will be held.

Written Submissions: In lieu of or in addition to participating in the hearing, interested parties are invited to submit written statements concerning this investigation. All written submissions should be addressed to the Secretary, and should be received not later than 5:15 p.m., May 20, 2012. All written submissions must conform with the provisions of section 201.8 of the *Commission's Rules of Practice and Procedure* (19 CFR 201.8). Section 201.8 requires that a signed original (or a copy so designated) and fourteen (14) copies of each document be filed. In the event that confidential treatment of a document is requested, at least four (4) additional copies must be filed, in which the confidential information must be deleted (see the following paragraph for further information regarding confidential business information). The Commission's rules authorize filing submissions with the Secretary by facsimile or electronic means only to the extent permitted by section 201.8 of the rules (see Handbook on Electronic Filing Procedures, *http:// www.usitc.gov /docket_services/documents /handbook_on_electronic_filing.pdf*).

Persons with questions regarding electronic filing should contact the Secretary (202–205–2000).

Any submissions that contain confidential business information (CBI) must also conform with the requirements of section 201.6 of the *Commission's Rules of Practice and Procedure* (19 CFR 201.6). Section 201.6 of the rules requires that the cover of the document and the individual pages be clearly marked as to whether they are the "confidential" or "non-confidential" version, and that the confidential business information be clearly identified by means of brackets. All written submissions, except for confidential business information, will be made available for inspection by interested parties.

In his request letter, the USTR stated that he intends to make the Commission's report available to the public in its entirety, and asked that the Commission not include any confidential business information or national security classified information in the report. Any confidential business information received by the Commission in this investigation and used in preparing this report will not be published in a manner that would reveal the operations of the firm supplying the information.

By order of the Commission.

Issued: July 20, 2011.

James R. Holbein,

Secretary to the Commission.

APPENDIX C: HEARING PARTICIPANTS

Calendar of Public Hearing

Those listed below appeared as witnesses at the United States International Trade Commission's hearing:

Subject: Remanufactured Goods: An Overview of the U.S. and Global Industries, Markets, and Trade

Inv. No.: 332-525
Date and Time: February 28, 2012 - 9:00 a.m.

Sessions were held in connection with this investigation in the Main Hearing Room (room 101), 500 E Street, S.W., Washington, DC.

Organization and Witness

Panel 1

Rochester Institute of Technology ("RIT")
Golisano Institute for Sustainability
Rochester, NY

Nabil Z. Nasr, Ph.D.,
Assistant Provost of Academic Affairs and Director,
Golisano Institute for Sustainability and CIMS
Remanufacturing Industries Council ("RIC") Rochester, NY
Matthew P. Hehl, Executive Director
Motor & Equipment Manufacturers Association ("MEMA")
Washington, DC
Robert McKenna, President and CEO
Motor & Equipment Manufacturers Association ("MEMA")
Washington, DC
Rob Wright, Managing Director, Undercar Express (UCX)
Caterpillar Inc. Mossville, IL
Greg Folley, Vice President,
Remanufacturing Goods & Components Division

Organization and Witness

Panel 2

Detroit Diesel Remanufacturing LLC Detroit, MI
Brian A. Lewallen, Operations Director
Cummins Inc. Washington, DC
Allen W. Pierce, General Manager, New and Recon Parts Manufacturing
John Deere Reman Springfield, MO
Donald R. Flatau, General Manager, Global Reman,
Operations & Marketing
MacKay & Company Lombard, IL
David Fulghum, Vice President
Tire Industry Association ("TIA") Bowie, MD
Marvin Bozarth, Senior Technical Consultant
Tire Retread & Repair Information Bureau ("TRIB") Falls Church, VA
David Stevens, Managing Director

International Imaging Technology Council ("IITC") Las Vegas, NV
Tricia Judge, Executive Director

Organization and Witness

Panel 3

Davies Office Reburbishing, Inc. Albany, NY
William E. Davies, President
GE Healthcare Systems Waukesha, WI
Michael R. Schmit, Remanufacturing Technology Director
Siemens Medical Solutions USA, Inc. Malvern, PA
Hans Beinke, Vice President, Healthcare,
US Country Head, Quality and Technology
Association of Medical Device Reprocessors ("AMDR") Washington, DC
Daniel J. Vukelich, Esq., President
PC Rebuilders and Recyclers ("PCRR") Chicago, IL
Willie Cade, Owner
Reverse Logistics Association Lehi, UT
Jesse LaRose, Member

APPENDIX D: POSITIONS OF INTERESTED PARTIES

Summary of Positions of Interested Parties

The U.S. International Trade Commission (USITC or Commission) held a public hearing in relation to its investigation on the global industries, markets, and trade in remanufactured goods on February 28, 2012, in Washington, DC. Interested persons were also invited to file written submissions for the investigation. This appendix contains a summary of the views expressed to the Commission via testimony, written submission, or both, and reflects the principal points made by the particular party. The views expressed in the summarized materials should be considered to be those of the submitting parties and not of the Commission. In preparing this summary, Commission staff did not undertake to confirm the accuracy of, or otherwise correct, the information summarized. For the full text of hearing testimony, written submissions, and exhibits, see entries associated with investigation No. 332-525 at the Commission's Electronic Docket Information System (https://edis.usitc.gov /edis3-internal/app).[1]

Association of Medical Device Reprocessors (AMDR)[2]

Mr. Daniel Vukelich, president of AMDR, noted that the organization is a trade association that represents the legal, regulatory, and legislative interests of the U.S.- based, and U.S. Food and Drug Administration (FDA)-regulated, third-party reprocessors of medical devices labeled by the original manufacturer as "single use." Mr. Vukelich stated that AMDR estimates that its members perform approximately 95 percent of the third-party reprocessing of medical devices labeled as "single use" in the United States. He said that the U.S. market for reprocessed medical devices is estimated at $250 million annually, and is anticipated to grow by 12–25 percent through 2013.

According to Mr. Vukelich, the FDA considers single-use medical device reprocessors, such as AMDR's members, to be manufacturers of single-use medical devices. He said that medical device reprocessors contract with hospitals and surgical centers to collect previously used medical devices, which are labeled by their original equipment manufacturer as "single use," and then reprocess them for another "single use" by using scientifically supported, FDA-accepted methods.

Mr. Vukelich said that since 2000, the FDA has considered third-party reprocessors of single-use devices to be manufacturers and, therefore, subject to FDA medical device manufacturer requirements. He also said that by complying with FDA regulatory requirements, reprocessors offer a product that is as safe and effective as a new device and poses no increased risk to patients. Moreover, Mr. Vukelich stated that the "single-use" label is a designation chosen by medical device manufacturers, and not by the FDA.

According to Mr. Vukelich, FDA-regulated reprocessed devices cost about one-half of what an original device costs. Mr. Vukelich said that reprocessors estimate that a typical 200-bed hospital, if purchasing a reprocessor's full product line instead of new products, can save between $600,000 and $1 million per year and divert 5,000–15,000 pounds of waste from landfills.

Mr. Vukelich said that internationally, the reprocessing of single-use devices is considered to be commonplace, but that it is seldom regulated. He said that Canada and the European Union (EU) present significant market opportunities for U.S. third-party reprocessors, but that neither has a comprehensive regulatory framework regarding the processing of single-use devices. Mr. Vukelich said that Canada does not have any federal regulatory requirements for reprocessors of single-use devices, but that reprocessing is regulated at the provincial level. As a result, he said, some provinces allow hospitals to contract with U.S. third-party firms, while other provinces do not. Mr. Vukelich said that while the EU does not have a comprehensive regulatory framework regarding the reprocessing of single-use devices, it is in the process of revising its medical device regulatory framework to include provisions that will more thoroughly address reprocessed medical devices on the European market.

Automotive Parts Remanufacturing Association (APRA)[3]

In a written submission to the Commission, Mr. William Gager, president of APRA, said that APRA has represented the automotive and truck parts remanufacturing industry since October 1941. He estimates that there are more than 15,000 companies in the United States that remanufacture motor vehicle parts. Mr. Gager said that there are many large companies employing between 100 and over 4,000 people who remanufacture a wide variety of motor vehicle parts. Likewise, Mr. Gager stated that there are thousands of smaller companies that employ fewer than 20 employees that perform some type of remanufacturing activity on a variety of parts. He also estimates that the number of employees in motor vehicle parts remanufacturing in the United States may exceed 350,000.

Mr. Gager said that the U.S. market for remanufactured goods is strong and that trade in remanufactured goods is important to the U.S. economy. He said that as passenger car ownership grows in China, India, and other parts of the world, the remanufacturing industries will grow as a result of these countries figuring out how to handle repairs of their growing number of passenger cars.

Mr. Gager concluded by stating that APRA would support the gathering of remanufacturing-specific industry information by the U.S. government, for such information would help APRA members to determine market size and growth opportunities. Mr. Gager also stated that APRA hopes that the U.S. Department of Commerce (USDOC) will continue to break down barriers to trade in remanufactured products.

Caterpillar Inc. (Caterpillar)[4]

In written and oral testimony to the Commission, Mr. Greg Folley, vice president of Caterpillar's remanufacturing goods and components division, said that Caterpillar is a global leader in remanufacturing and that remanufacturing is an important part of its business. He stated that Caterpillar and its subsidiaries—Electro-Motive Diesel, Progress Rail Services, and Solar Turbines—offer a wide range of remanufactured goods and services. Mr. Folley stated that these remanufactured products range from engines, turbines, gas compressors, locomotives, and railcars to hydraulics, drivetrains, turbochargers, and fuel systems. Mr. Folley said that Caterpillar's remanufacturing activities employ more than 8,000 people in 68 plants in 35 U.S. states and in 15 countries, and that Caterpillar remanufactures more than 14,500 products for its own business entities and for outside customers.

Mr. Folley said that remanufacturing is the process of returning end-of-life products to the same as their original, new condition or better in a manufacturing environment. He noted that remanufacturing benefits the environment by reducing the amount of material going into a landfill, using considerably less energy than manufacturing a new product, and reducing the demand for raw materials to make new products.

According to Mr. Folley, Caterpillar's remanufactured products are sold in 170 countries and in most countries remanufactured products are freely traded. He noted, however, that in many countries—particularly Brazil, China, India, and Russia—nontariff barriers and tariffs restrain Caterpillar's ability to freely trade remanufactured goods and cores. Mr. Folley stated that one nontariff barrier is the absence of a definition of remanufactured products, resulting in customs officials in some countries treating remanufactured products as used products. Mr. Folley noted that some countries often ban the importation of used products despite the fact that Caterpillar's remanufactured products are comparable in quality and warranty coverage to those of new products. Mr. Folley said that countries should treat remanufactured goods in the same way they treat new goods.

Mr. Folley said that some countries restrict the movement of cores, which are vital for the operations of remanufacturers. He stated that stringent export documentation requirements can make it difficult or impossible for a customer to export a core. Mr. Folley said that imprecise or underdeveloped tax laws in some countries can have a negative effect on remanufacturers and their customers.

Mr. Folley concluded by stating that Caterpillar's remanufactured products are good for its customers, good for its business, and good for the environment.

Cummins Inc. (Cummins)[5]

In written and oral testimony to the Commission, Mr. Allen Pierce, general manager of new and recon parts manufacturing for Cummins Inc., said that Cummins is a global company that designs, manufactures, distributes, and services engines and related technologies, including fuel systems, controls, air handling, filtration, emission solutions, and electrical power generation systems. He stated that Cummins is headquartered in Columbus, Indiana, and serves customers in about 190 countries and territories. Mr. Pierce said that the Cummins remanufacturing business began more than 45 years ago, with annual sales approaching nearly $1 billion in 2011, which accounted for approximately 5.5 percent of Cummins's total corporate annual sales in that year. He noted that global employment in the Cummins remanufacturing business totaled 2,843 at year end 2011, and is projected to grow to 3,206 by the end of 2012. Mr. Pierce stated that Cummins has two remanufacturing plants in the United States and eight remanufacturing plants in foreign countries. He said that Cummins anticipates continued growth for its remanufactured products and is making the necessary global investments to meet that demand.

According to Mr. Pierce, remanufacturing involves completely disassembling the product, cleaning and inspecting each individual component, remanufacturing the parts to same as new performance, testing the remanufactured product to ensure it meets the same quality standards as a newly manufactured product, and then providing the same warranty for the remanufactured product as for a new product. He said that remanufacturing enables Cummins to be a good steward of the environment, provides cost savings of 20– 40 percent to its customers compared with purchasing a new product, and meets its customers' growing demands for sustainable products.

Mr. Pierce stated that Cummins sees opportunities for growth in remanufacturing in foreign countries, but it is also concerned about trade barriers to remanufactured products in some of these countries. According to Mr. Pierce, there is a perception in these countries that remanufactured products are old, defective, and used products, rather than products that are like new and that meet all of the same quality standards as newly manufactured products. Mr. Pierce stated that Cummins has encountered various types of trade barriers to its remanufacturing activities in Argentina, Belarus, Brazil, Chile, China, Colombia, India, Kazakhstan, Russia, and Venezuela. Mr. Pierce said that the growth of remanufacturing worldwide is being constrained by the lack of a common understanding of the industry. He stated that Cummins recommends that the remanufacturing industry and government collaborate in developing a common definition of and global standards for remanufacturing, which would help remove trade barriers that restrict the free movement of cores and remanufactured products. Mr. Pierce stated that Cummins would like to see national restrictions removed so that it can provide remanufactured goods as repair and refurbishment options for Cummins-powered equipment in all countries. He said that Cummins would prefer that remanufactured goods be treated the same way new goods are treated and encourages U.S. government support in this regard.

Davies Office Refurbishing Inc. (Davies)[6]

In oral testimony before the Commission, Mr. William Davies, president of Davies, stated that his company has been in business since 1948, currently has a 300,000-squarefoot manufacturing facility, and employs 160 full-time workers. Mr. Davies said that office furniture remanufacturers engage in four processes: metal finishing, laminate work and millwork, wood finishing and refinishing, and upholstery. According to Mr. Davies, the remanufacturing of office furniture in the United States began in the late 1980s and in 1990 accounted for roughly $325,000 in sales out of a total U.S. marketplace for newly manufactured office furniture and remanufactured office furniture of $8.5 billion. Mr. Davies said that currently, sales of remanufactured office furniture come to $1.4 billion out of a total $10.4 billion U.S. market for newly manufactured and remanufactured office furniture. Mr. Davies stated that remanufactured office furniture in general saves the purchaser 30 to 70 percent over newly manufactured office furniture.

According to Mr. Davies, the remanufactured office furniture industry is fragmented and consists of recyclers, remanufacturers, refurbishers, and brokers. He noted that in 1990 there were approximately 59 firms engaged in the remanufacture of office furniture, and they employed more than 2,500 full-time workers. Mr. Davies stated that in 2009 there were roughly 50 firms involved in the remanufacture of office furniture, and so the overall number of firms in the industry has not changed that much over the past two decades.

Finally, Mr. Davies noted that the remanufactured office furniture industry has dealt with the ups and downs of the real estate market over the years and that it has faced increased import competition in recent years. He stated that the industry has evolved in order to compete with the manufacturers of new office furniture, and that it has begun to work with these manufacturers to develop common standards.

Detroit Diesel Remanufacturing (DDR)[7]

In written and oral testimony to the Commission, Mr. Brian Lewallen, operations director at DDR, said that DDR is a remanufacturer of products for on- and off-highway power systems and is part of Daimler Trucks North America, a leading manufacturer of class 6– 8 commercial trucks in the United States. He said that DDR has operated as a U.S. remanufacturer since 1966 and currently employs 1,217 workers in six facilities located in Kansas, Michigan, Minnesota, Ohio, and Utah.

According to Mr. Lewallen, remanufacturing occurs in many industries, including the home appliances, furniture, medical devices, electronics, aviation, automotive, and trucking industries. He said that remanufacturing is recognized as the ultimate form of recycling because remanufacturing provides a closed-loop process that renovates products that are at the end of their useful lives, thus minimizing their impact on the environment and landfills. Mr. Lewallen said that the remanufacturing process saves up to 85 percent of the energy and material used to manufacture equivalent new parts, reduces carbon dioxide emissions by 53 percent, saves 14 million tons of raw materials annually, and generates 75 percent less landfill waste.

Moreover, Mr. Lewallen said that remanufactured engines and parts cost on average 20–30 percent less than new engines and parts, and that the potential cost savings from producing remanufactured products instead of new goods are between 30 and 70 percent. He also said that it is estimated that the U.S. remanufacturing sector employs over 500,000 workers.

Mr. Lewallen noted that the U.S. remanufacturing industry faces several obstacles, including a lack of consumer understanding and awareness, a lack of worldwide standards, nontariff barriers in foreign countries, and inconsistent use of remanufactured products. He said that remanufactured products are sometimes misunderstood both domestically and internationally, and are considered used products in some countries. He also said that due to a lack of worldwide standards, the difference between remanufactured goods and used or rebuilt products is not defined, resulting in confusion in foreign markets as to how cores and products are managed. Moreover, Mr. Lewallen said that nontariff barriers in certain countries limit U.S. exports of remanufactured goods. More specifically, he said that Argentina, Brazil, Colombia, Ecuador, and Peru restrict the free movement of remanufactured goods and cores. In the United States, Mr. Lewallen said that U.S. government entities, including local municipalities and the U.S. armed services, do not endorse the use of remanufactured goods to maintain equipment and vehicles.

Dr. Nabil Nasr, Assistant Provost of Academic Affairs and Director, Golisano Institute for Sustainability, Rochester Institute of Technology (RIT)[8]

In hearing testimony before the Commission, Dr. Nasr provided his views on why he thinks remanufacturing is important to industrial economies around the world. He described a closed-loop industrial system in which goods that have reached the end of their useful lives are recovered to ensure proper disposal of the material and components of the products. Dr. Nasr said that remanufacturing is the most economical and environmentally benign way to recover, restore, and reuse these products in multiple cycles in order to reduce pollution and address material shortages in industrial economies. According to Dr. Nasr, remanufacturing helps minimize waste and conserve energy, and provides significant opportunity for employment in the United States. Dr. Nasr noted that the cost of remanufactured goods is typically 25–50 percent lower than that of new goods.

Dr. Nasr stated that the United States is the largest remanufacturer in the world, and that remanufacturing occurs in all 50 states. Dr. Nasr said that the U.S. Department of Defense is considered to be the largest remanufacturer in the world; examples of products that it remanufactures include aircraft, ground vehicles, and weapons systems.[9] He said that the automotive sector is the largest remanufacturer in the commercial sector. According to Dr. Nasr, other principal remanufacturing sectors in the United States are aerospace, consumer products, electrical apparatus, furniture, heavy-duty and construction equipment, imaging products, information technology equipment, locomotive systems, machinery, medical equipment, restaurant equipment, and tires.

Finally, Dr. Nasr noted that North American Industry Classification System (NAICS) codes do not distinguish between remanufactured and new products, and as a result, industry information for remanufactured products is unavailable. He said that the Remanufacturing Industries Council (RIC) had advocated for the creation of separate NAICS codes in order to help the remanufacturing industry identify the size and scope of remanufacturing in the United States.[10]

Electrical Apparatus Service Association (EASA)[11]

In a written submission to the Commission, Ms. Linda Raynes, president and CEO of EASA, a trade association representing the electromechanical equipment repair and sales industry, stated that EASA considers the definition of remanufactured goods used by the Commission in its investigation to be too broad and possibly misleading. EASA stated that the definition fails to distinguish between "remanufacturing" and "rebuilding." According to EASA, its members are rebuilders and not remanufacturers, and the Commission should modify its definition so that rebuilders are not considered remanufacturers.

EASA stated that in remanufacturing, cores and parts from many different machines are used to create a new machine, whereas in rebuilding, an existing machine uses only the original machine plus the parts needed to restore that machine to "like-new or better condition." EASA stated that its members do not remanufacture motors and generators in quantity; instead, its members rebuild motors and generators on a single-replacement basis. According to EASA, the term "disassembly" in the Commission's definition of remanufactured goods does not accurately describe the rebuilding process, as motors that are rebuilt are only partially disassembled. Finally, EASA noted that remanufacturers take ownership of the unit and sell the remanufactured product. In contrast, rebuilders do not take ownership of the product; rather, rebuilders sell to a customer the parts and labor that are needed to put the unit in a condition that equals or betters the original manufacture's specifications. EASA noted that in this case, the customer retains ownership of the unit.

The Association of Electrical Equipment and Medical Imaging Manufacturers (NEMA)[12]

In a written submission to the Commission, Mr. Kyle Pistor, vice president, government relations of NEMA, wrote that NEMA is the association of electrical equipment manufacturers whose member companies manufacture a diverse set of products, including power transmission and distribution equipment, lighting systems, factory automation and control systems, and medical diagnostic imaging systems. In addition,

Mr. Pistor said that the Medical Imaging Technology Alliance (MITA), a division of NEMA, represents companies whose sales comprise more than 90 percent of the global market for medical imaging technology.

According to Mr. Pistor, the current world market for remanufactured medical imaging equipment is estimated to be $1.5 billion annually. He said that the U.S. market for remanufactured medical equipment is estimated to be approximately $560 million annually, and that U.S. remanufacturers supply 90 percent of the U.S. market for remanufactured medical equipment. Mr. Pistor said that medical imaging equipment is remanufactured by four major original equipment manufacturers (OEMs), as well as by many other smaller businesses. Mr. Pistor stated that remanufactured medical imaging equipment carries a full service warranty and support services just as a new product does.

Mr. Pistor said that declining reimbursement rates for medical imaging procedures are driving demand for remanufactured units from less financially well-endowed medical facilities. He said that a remanufactured unit could be considerably more affordable for smaller or rural medical facilities.

With respect to trade in remanufactured medical imaging equipment, Mr. Pistor said that significant import barriers exist. Mr. Pistor said that China, Egypt, Kuwait, Syria, and Thailand do not allow the importation of any remanufactured medical imaging equipment. He also said that 24 countries restrict imports of remanufactured medical equipment. Mr. Pistor said that measures include high tariffs, special certification requirements, equipment age restrictions, and bans on competing with locally produced devices.

Mr. Pistor added that OEM manufacturers have also invested in remanufacturing plants outside of the United States, such as in India, because some countries only allow the sale of remanufactured equipment in their market if it was remanufactured there. In addition, he said that because the cores of medical imaging devices are large, making it impractical to ship them over long distances, many remanufactured imaging devices are manufactured within the country in which the cores are sourced. He said that up to 80 percent or more of cores are sourced from the local market.

For other types of electrical equipment, Mr. Pistor said that electric motors, motor drives, some industrial process controls, and electric transformers are also remanufactured or refurbished. He said that in these industries there are fine distinctions between repair, refurbishing, and remanufacturing. For motors, he said, rewinding is often referred to as remanufacturing.

Finally, regarding trade in used goods, Mr. Pistor stated that ongoing negotiations under the auspices of the Basel Convention on the Control of Transboundary Movements of Hazardous Wastes and Their Disposal are nearing a decision to extend export controls and trade prohibitions to cover shipments of used products for repair, refurbishment, and reuse. Mr. Pistor said that the EU proposals would apply hazardous-waste controls and trade bans to exports of most used electrical and electronic equipment. As a result, Mr. Pistor said, manufacturers that rely on the transboundary movement of legitimate equipment for legitimate service, repair, refurbishing, and remanufacturing would face massive costs.

GE Healthcare Systems (GE)[13]

In written and oral testimony to the Commission, Mr. Michael Schmit, remanufacturing technology director for GE, stated that the most common medical equipment that is remanufactured includes magnetic resonance, computed tomography, X-ray, interventional cardiology, positron emission tomography/computed topography (PET/CT), and nuclear and ultrasound imaging devices. Mr. Schmit said that U.S. producers of this remanufactured equipment are the four original equipment manufacturers (GE, Siemens, Phillips, and Toshiba) and more than 100 small to medium-sized businesses. He stated that sales of remanufactured medical imaging equipment totaled $564 million in the United States in 2010, with sales expected to grow 5 percent compounded annually through 2015. Mr. Schmit said that the major customers for this equipment are hospitals, imaging centers, and private practices.

According to Mr. Schmit, remanufacturing of medical imaging equipment expands the affordability of healthcare by allowing a customer with an original system to trade it in after 3 to 5 years, either because the term of the lease has expired or the customer wants to upgrade to the latest technology. He stated that the customer trades in the system and uses the trade-in payment to help buy the newest equipment. Mr. Schmit noted that the returned system is remanufactured and recertified to the original specification, is backed by a full-service warranty and full-service support, and is then sold to a customer at a significantly reduced price.

Mr. Schmit stated that the current trends in the United States for healthcare appear to be favorable for the remanufactured medical imaging equipment industry. He said that declining reimbursements for imaging procedures make lower-priced remanufactured medical imaging equipment more attractive to customers. He also noted that technology innovation leads to quick turnover of medical imaging equipment, which provides the material for remanufacturing.

Finally, Mr. Schmit said that a U.S. Department of Commerce publication indicates that many countries restrict the importation of remanufactured medical imaging equipment. He noted that these restrictions include outright bans, high tariffs or fees, bans on use by public institutions, special certification requirements, excessive documentation, equipment age restrictions, and a ban on equipment that competes with locally produced devices.

International Imaging Technology Council (IITC)[14]

In written and oral testimony to the Commission, Ms. Tricia Judge, executive director of IITC, said that the IITC is a trade association representing 2,500 North American companies that remanufacture imaging supplies such as toner cartridges and ink jet cartridges. She stated that a "remanufactured printer cartridge is a cartridge that has been restored to its original performance standards and function and is thereby diverted from the solid waste stream, retaining, to the extent practicable, components that have been through at least one life cycle and replacing consumable or normal wear components."

According to Ms. Judge, remanufactured cartridges are beneficial in three ways: (1) they reduce the amount of plastic disposed in landfills, thus conserving the oil that would have been used in producing a new cartridge; (2) the remanufacture of cartridges is labor intensive, so cartridge remanufacturers provide jobs for local communities; and (3) remanufactured cartridges are a lower-price alternative to new cartridges. She also noted that over the years the competition between OEMs and the remanufacturers has been intense.

Ms. Judge noted that toner and ink jet cartridge remanufacturing started in the mid-1980s, when individuals discovered that these cartridges could be replenished with toner and reused. Ms. Judge said that over the years, the quality of remanufactured cartridges has increased to the point that they are now sold alongside new cartridges in the stores of retailers and through catalogs and shops worldwide. Ms. Judge said that the U.S. remanufactured printer cartridge industry grew significantly, such that at its peak it consisted of 6,000 companies with sales of $8 billion and a one-third share of the total cartridge market.

According to Ms. Judge, the U.S. cartridge remanufacturing industry at present consists of 2,000 companies employing 33,650 workers and sales of $3.5 billion, down considerably from the 6,000 companies and sales of $8 billion at the industry's peak some years ago. Ms. Judge noted that this decline is due to a number of factors. She stated that there have been core shortages caused by the OEMs' aggressive core destruction programs, which the OEMs describe as "recycling" efforts. Ms. Judge stated that the weak economy has caused higher unemployment, which has lowered demand for printers. Ms. Judge said that increased imports, primarily from China, of "new compatible" or "new remanufactured" cartridges are displacing U.S. remanufactured cartridges. According to Ms. Judge, these imported cartridges are often counterfeit, of inferior quality, and sold at prices far below what U.S. remanufacturers spend to remanufacture cartridges. Ms. Judge said that U.S. cartridge remanufacturers' share of world remanufactured cartridge production has declined from one-half to one-third.

John Deere Reman[15]

In written and oral testimony to the Commission, Mr. Donald Flatau, general manager of global remanufacturing operations and marketing for John Deere Reman, said that John Deere Reman has its global headquarters in Springfield, MO, and additional operations in Edmonton, Alberta, Canada. Mr. Flatau said that John Deere Reman offers a vast portfolio of remanufactured products, including off-road diesel engines, construction and forestry machines, and components such as starters, alternators, turbochargers, fuel injection pumps, transmissions, and axles. According to Mr. Flatau, John Deere Reman has enjoyed significant growth in recent years, with revenue more than quadrupling since 2003, from $45 million to $200 million, and employment increasing by 11 percent in the past 12 months.

Mr. Flatau stated that remanufacturing benefits the environment by reusing components that otherwise would have ended up in a landfill and reduces the sector's carbon footprint because remanufacturing uses less energy than manufacturing a new product. Mr. Flatau said that John Deere Reman's Springfield facility remanufactured more than nine million pounds of components in 2011, which otherwise would have been discarded.

However, Mr. Flatau noted that one of the challenges facing the remanufacturing industry is clearly defining remanufacturing and agreeing to industry quality standards. He stated that John Deere Reman guarantees that its remanufactured products are built to the same quality standards as that of its newly manufactured products. Mr. Flatau stated that the warranty on these remanufactured products is a genuine John Deere warranty that meets or exceeds the coverage level found with the equivalent new product. He noted that in the John Deere Reman remanufacturing process the products are completely disassembled and cleaned; all parts are thoroughly inspected and either replaced or salvaged by using appropriate reclamation techniques, then reassembled to original specifications or better; and finally the products are fully tested to original specifications or better.

According to Mr. Flatau, the remanufacturing industry needs independent NAICS codes to define and measure it. He noted that the rest of the world looks to the United States for leadership in remanufacturing, and yet it is not classified as a separate industry here. He stated that setting industry standards for remanufacturing is important for the industry to be recognized worldwide for the economic driver that it is.

Finally, Mr. Flatau said that another challenge facing John Deere Reman's business is trade barriers to its remanufactured products in many countries. He stated that demand for these products is large and growing in countries such as Brazil, India, Argentina, China, and Turkey, but that customers there are prohibited from buying them. Mr. Flatau said that in many countries, remanufactured products are considered used, dirty, or inferior, and thus barriers have been imposed to keep them out. Mr. Flatau said that once remanufacturing standards are clearly defined, this must be communicated and considered in trade agreement negotiations. He noted that the ability of John Deere Reman to grow its business is dependent upon gaining access to these rapidly growing foreign markets.

MacKay & Company (MacKay)[16]

In written and oral testimony to the Commission, Mr. David Fulghum, vice president of MacKay, said that his company is a small research business consulting firm located in the suburbs of Chicago, IL. Mr. Fulghum stated that the company's area of research is the commercial vehicle market, which includes commercial on-highway vehicles (Class 2c8), construction equipment, farm field machinery, and related parts and components such as engines, transmissions, and tires. He noted that MacKay collects data through an annual survey of thousands of end users on their aftermarket parts and service purchase practices. Mr. Fulghum said that the survey process is supported by a large database of owners and maintainers of commercial trucks, farm field equipment, and off-highway construction equipment and by a large database of parts and service suppliers. Mr. Fulghum stated that the firm's data analysis is based on end user purchases and is valued at current retail prices. He noted that the firm's database and the data presented in his testimony are not all-inclusive with respect to the commercial vehicle market.

According to Mr. Fulghum, the value of remanufactured parts sold into the commercial vehicle market in the United States in 2011 was an estimated $15.2 billion. He stated that 84 percent of this value was accounted for by remanufactured parts sold for commercial trucks, 9 percent by parts sold for construction equipment, and 7 percent by parts sold for farm field equipment. Finally, he noted that other data also show that most remanufactured parts are sold into the commercial vehicle market in the United States.

Medtronic Inc. (Medtronic)[17]

In a written submission to the Commission, Mr. Michael Morton, senior director for global regulatory affairs at Medtronic, stated that his company is a global leader in medical technology, and that Medtronic develops and manufactures a wide range of products and therapies, including single-use medical devices. Mr. Morton said that Medtronic also remanufactures a small number of medical devices.

Mr. Morton stated that medical devices that are intended to be reused are designed so that they can be reprocessed or remanufactured to their original performance specifications. In contrast, Mr. Morton noted that devices that are designed and labeled for "single-use only" cannot be assured to meet original performance specifications after reprocessing.

Mr. Morton said that as a manufacturer of single-use medical devices, Medtronic considers the practice of reprocessing and marketing single-use devices to pose significant risks for patients, yielding products and that may be neither safe nor effective. According to Mr. Morton, medical devices labeled for single use have been designed and materials have been selected to perform for its intended use, and not for cleaning and/or sterilization. He said that since many reprocessors are not associated with the original manufacturers, the reprocessors likely do not know the original specifications. Mr. Morton stated that Medtronic therefore contends that the reprocessing of single-use devices should not be encouraged.

Motor and Equipment Manufacturers Association (MEMA)[18]

In written and oral testimony to the Commission, Mr. Robert McKenna, president and chief operating officer of MEMA, said that MEMA represents over 700 companies that manufacture and remanufacture motor vehicle parts for use in the light- and heavy-duty original equipment and aftermarket industries. He said that MEMA represents its members through four affiliate associations: Automotive Aftermarket Suppliers Association (AASA), Heavy Duty Manufacturers Association (HDMA), Motor and Equipment Remanufacturers Association (MERA),[19] and Original Equipment Suppliers Association (OESA).

Mr. McKenna said that the U.S. automotive aftermarket (non-heavy duty) consists of over 240 million passenger cars and $185 billion in annual business, with annual parts sales of $78 billion. He said that remanufactured parts sales account for 25 percent, or approximately $20 billion, of U.S. aftermarket sales. Mr. McKenna said that 3,800 firms employ over 350,000 workers in the motor vehicle remanufacturing sector. According to Mr. McKenna, remanufactured motor vehicle parts on average cost 20–50 percent less than most new parts. In a posthearing written submission, MEMA stated that the average cost savings on remanufacturing products compared with new aftermarket products is 40 percent.[20]

Mr. McKenna also noted that in 2011, MEMA created an affiliate organization focused on remanufacturing—MERA—in response to the growing recognition of the important role that remanufacturing has in the United States and global economies. According to Mr. McKenna, among MERA's top priorities is an education and awareness campaign to heighten the understanding and acceptance of remanufacturing. He said that the campaign will include a certification program that will inform customers that remanufactured parts from MERA member companies are truly remanufactured. He said that the program will define minimum process criteria for products to be eligible for certification. In a posthearing written submission, MEMA stated that it considers the definition of remanufactured goods used by the Commission in its investigation to be suitable, and that the definition will help promote trade in remanufactured goods.[21]

With respect to trade in remanufactured goods, Mr. McKenna said that trade barriers—particularly nontariff barriers—are one of the key obstacles to growing the remanufacturing industry. As an example, he noted that the North American Free Trade Agreement (NAFTA) did not provide for cores to meet the rules of origin granting duty-free treatment. Although the United States and Mexico subsequently made the necessary changes in law to allow cores to receive duty-free treatment, he said, Mexico has yet to implement the changes in practice. Mr. McKenna said that this is important because 75 percent of U.S. exports of motor vehicle parts are to Canada and Mexico. Mr. McKenna said that MEMA and MERA continue to work with the Office of the United States Trade Representative (USTR) and DOC to promote remanufactured goods trade with U.S. trading partners, including those engaged in Trans-Pacific Partnership (TPP) trade negotiations and at the Asia-Pacific Economic Cooperation (APEC) forum.

PC Rebuilders and Recyclers (PCRR)[22]

In oral testimony to the Commission, Mr. Willie Cade, owner of PCRR, stated that the refurbishment of personal computers (PCs) involves the replacement of the operating system after the data on the hard drive have been destroyed. He noted that the data destruction is a critical element when equipment is transferred from one user to another, and that installing a new operating system is not an easy task. Mr. Cade said that PCRR is a remanufacturer (refurbisher) because it disassembles PCs, then processes, cleans, inspects, and tests them. He said that PCRR's products are better than new and have a warranty.

Mr. Cade stated that PCRR is one of 16 Microsoft-authorized refurbishers in the United States and that there are 76 Microsoft-authorized refurbishers worldwide. Mr. Cade said that Microsoft has a registered refurbisher program; there are about 1,200 registered refurbishers in the United States and approximately 2,800 globally. He stated that the refurbishment program is a contractual relationship which allows for the use of licensed software, an important feature for PCRR customers who want genuine product to ensure that they can continue to obtain monthly upgrades. Mr. Cade noted that much of the global demand for PCs will be met by refurbished or remanufactured PCs, not by new ones. He stated that a major concern of PCRR's customers is the destruction of data on their hard drives to ensure that their data are not released to the next purchaser of their refurbished computer. Finally, Mr. Cade said that to assure its customers in this regard PCRR follows the NIST-888 special publication on data destruction, which involves destroying data contained on customers' hard drives a traceable manner.

Remanufacturing Industries Council (RIC)[23]

Mr. Matthew Hehl, executive director of RIC, said that the organization has 16 member companies representing a diverse group of remanufacturing sectors. In his testimony, Mr. Hehl said that RIC was established to cover problems faced in common by numerous and diverse remanufacturing sectors, and by design is not sector-specific. He emphasized that remanufacturing is a controlled process that adds jobs, promotes sustainability of resources, supports the environment, and enhances growth and innovation through many sectors of the economy. Mr. Hehl said that the processes employed by remanufacturers result in end products that are as good as, if not better than, the original. He also said that the process of remanufacturing prolongs the economic value of a product over its life cycle, thereby increasing the return on invested capital for businesses of all sizes.

Also in his testimony, Mr. Hehl stated several reasons why remanufacturing is a key component of the broader manufacturing ecosystem. In addition to the cost and environmental benefits of remanufacturing, Mr. Hehl noted that remanufacturing helps identify design problems in new products and production processes. In addition, Mr. Hehl noted that remanufactured products help OEMs reduce their need to invest in new inventory to support after-market service requirements by eliminating the necessity of all time requirements.

Reverse Logistics Association (RLA)[24]

In written and oral testimony to the Commission, Mr. Jesse LaRose, a member of the RLA, said that the RLA was formed to bring together all of the players in the reverse supply chain, such as third-party processing companies, OEMs, and the service providers and consultants that support both of these groups. Mr. LaRose stated that the RLA's mission is to educate industries about reverse logistics processes, which include everything that happens to a product once it gets returned to the retailer or manufacturer, such as customer service, fulfillment services, recycling, warranty replacement, refurbishment, and information technology process management. Mr. LaRose said that members of the RLA include many well-known brand name companies such as IBM, FedEx, Oracle, Dell, and Best Buy.

According to Mr. LaRose, remanufacturing in the United States is estimated to account for about 2.3 percent of GDP, that processing facilities number in the hundreds, and that there are thousands of garage shops which use online distribution channels. He noted that consumers are increasingly comfortable with purchasing refurbished products in place of new products, and thus large retailers are interested in capturing this growing market. Mr. LaRose stated that manufacturers are increasingly using refurbished goods as replacements for warranty returns, allowing manufacturers to have more control over their brands and to reduce costs and losses.

Mr. LaRose also said that remanufacturing is primarily a domestic activity in the United States serving the domestic market, although surplus product is exported when practical. However, he stated that exports of used goods are limited in some instances by punitive taxation by some countries. Mr. LaRose said that the consumer electronics refurbishment process involves a number of steps, including the return or take-back of the product to the manufacturer; the refurbishment of the product or the harvest of the product for its parts, with the remainder destined for scrap; and then the distribution of the refurbished product and the parts directly to the consumer, to manufacturers (as warranty replacements), to discount retailers, to online retailers, and to the export market.

Finally, Mr. LaRose stated that the RLA's goal is to be able to easily export refurbished goods that are surplus to the U.S. market, thus creating jobs in the United States, reducing the trade deficit, and keeping these products out of landfills. Mr. LaRose said that the RLA requests that the U.S. government, when negotiating trade agreements, look at reducing trade barriers not only for new goods but for used goods as well. Mr. LaRose stated that the RLA also requests that the USITC review existing trade agreements to clarify tax and duty requirements with respect to used goods and to renegotiate existing trade agreements to reduce barriers to U.S. exports of used goods.

Siemens Medical Solutions USA (Siemens)[25]

In written and oral testimony to the Commission, Mr. Hans Beinke, vice president of healthcare for Siemens, said that Siemens is a manufacturer of medical imaging devices such as X-ray, CT, radiation therapy, diagnostic ultrasound, molecular imaging, magnetic resonance, and remanufactured medical imaging devices. According to Mr. Beinke, the world market for remanufactured medical imaging equipment is approximately $1.5 billion annually, of which the U.S. market accounts for about $560 million. He noted that the U.S. medical imaging device remanufacturing industry consists of OEMs such as GE, Siemens, Philips, and Toshiba; and third-party remanufacturers. Mr. Beinke said that U.S. producers' share of the domestic market is about 90 percent.

Mr. Beinke stated that countries differ in their treatment of imported remanufactured medical imaging equipment. Some, such as China, do not allow imports of these products, while others, such as Japan, India, and Mexico, allow them. He noted that within most countries, purchasing rules for buying new versus remanufactured medical imaging equipment vary, depending on whether the purchaser is a public health provider, a private health provider, or the government.

According to Mr. Beinke, the recession and reimbursement cuts have required healthcare providers to look for medical imaging equipment that is both state of the art and economical. He said that remanufactured medical imaging equipment can be a solution for healthcare providers facing budgetary constraints; consequently, the market for this equipment is estimated to continue to grow 3 to 4 percent annually. Mr. Beinke stated that the main customers for remanufactured medical imaging equipment are smaller, rural hospitals and diagnostic imaging centers. He said that the availability of this remanufactured equipment is limited by the willingness of health care providers to replace their old equipment and thus make it available to be remanufactured.

Finally, Mr. Beinke stated that discussions are ongoing within the remanufactured medical imaging equipment industry on ways to harmonize terminologies and practices, thus facilitating the exchange of information globally and creating a common understanding of best practices within the industry. Mr. Beinke noted that the U.S. medical imaging device remanufacturing industry receives approximately $180 million annually in foreign direct investment. He said that U.S. OEMs have invested in remanufacturing plants in foreign countries, including India. Mr. Beinke stated that a reason for this investment is that because medical imaging device cores are large and bulky, remanufactured medical devices are typically remanufactured within the country where the cores are sourced.

Tire Industry Association (TIA)[26]

In oral testimony before the Commission, Mr. Marvin Bozarth, senior technical consultant for the TIA, said that the U.S. tire retread industry is one of the most successful recycling programs in the country today. Mr. Bozarth stated that retreading tires provides cost savings across the entire U.S. economy. Mr. Bozarth said that tire retreading began in the United States in 1904 but expanded greatly during World War II in response to the shortage of tires and the development of synthetic rubber. He stated that the U.S. military was and is a large consumer of retreaded tires and has even operated its own retreading plants. Mr.

Bozarth stated that retreaded tires are used on military and commercial aircraft, mining and construction vehicles, commercial trucks, school buses, and U.S. Postal Service mail delivery vehicles and larger mail transport trucks. He noted that retreaded tires cost less, sometimes significantly less, than new tires, and that retreaded tires have at times in the past met demand for tires that the new-tire industry was unable to meet.

According to Mr. Bozarth, retreaded tires are sold through the same distribution channels as new tires, with retreaders selling their retreaded tires through commercial truck tire centers, to retailers and wholesale dealers, and direct to trucking fleets, small trucking companies, and independent operators. Mr. Bozarth said that retreaded tires are serviced and warranted in basically the same way as new tires. Mr. Bozarth stated that tires frequently need repair for such things as nail holes and small chips, and thus tire repair has become an important part of the tire retreading industry. He noted that when tires can no longer be retreaded, they can be recycled into numerous types of other products, such as rubber mats and landscaping items.

Finally, Mr. Bozarth said that in the United States, tire retread plants are large multimillion-dollar investments. He said that in 2011 there were about 700 of these plants, with 18 of them producing large off-road tires, 5 producing aircraft retread tires, 4 producing passenger car retreads, and 680 producing truck retreads. Mr. Bozarth noted that many of these plants produce more than one type of retreaded tires. He stated that in 2011, 15.5 million truck retreads were produced in the United States.

Tire Retread & Repair Information Bureau (TRIB)[27]

In written and oral testimony to the Commission, Mr. David Stevens, managing director of TRIB, stated that TRIB is a trade association founded in 1974 with a mission to promote and defend tire retreading and tire repairing on a global basis. Mr. Stevens said that TRIB has more than 500 members, most of which are in the United States. Mr. Stevens noted that in addition to the cost savings of retreaded tires, the tire retread industry provides large environmental benefits. He said that retreading conserves oil—in 2011 the retreading of truck tires alone saved more than 232 million gallons of oil. Mr. Stevens said that retreading also provides significant savings in raw materials such as steel, natural rubber, and carbon black that are used in the production process. Mr. Stevens stated that retreading extends the life of tires and keeps millions of tires out of tire piles and landfills. Mr. Stevens also noted that the federal government supports the use of retreaded tires on government vehicles where appropriate. Moreover, Mr. Stevens stated that many state and federal studies have dispelled the myth that pieces of tire rubber commonly seen on the side of the road are from retreaded tires. Mr. Stevens noted in fact that these studies have determined that such tire fragments were from new and retreaded tires in equal proportion to their service on the roads and had virtually nothing to do with the processes used in the production of new tires or the retreading of tires. He stated that the two primary causes of tire damage identified were hitting something on the road and excessive heat on the tire.

Finally, Mr. Stevens said that retreaded tires are as safe, reliable, and long-lasting as new tires. He stated that the retreading industry continually strives to improve the quality of retreaded tires with advanced casing inspection technology, the use of shearography and X-ray machines, and the constant refinement and testing of rubber compounds and tread designs to provide better fuel economy and tread wear.

Undercar Express, LLC[28]

In written and oral testimony to the Commission, Mr. Rob Wright, managing director of Undercar Express, LLC, said that the company is a full-line remanufacturer of disk brake calipers in Cleveland, OH, that began business in 1997 and currently employs 62 workers.[29] Mr. Wright stated that the remanufacturers of end-of-life automotive parts to original-equipment or better specifications are primarily small, privately held businesses, such as Undercar Express. He also noted that collectively this industry in the United States has annual revenue of at least $20 billion, but that individually these businesses have annual revenue ranging from $10 million to $100 million.

According to Mr. Wright, remanufacturing is beneficial for the economy, the consumer, and the environment. He noted that remanufacturing is labor intensive, extends the useful life of consumer products, and provides products to consumers that are priced 25 to 50 percent below new products. Mr. Wright said that the remanufacture of goods saves approximately 85 percent of the energy that would have been used to make the same new good. Mr. Wright stated that remanufacturing also reduces the amount of carbon dioxide released into the atmosphere, reduces the amount of product going into landfills, and conserves large amounts of raw materials such as copper, iron, and aluminum.

Mr. Wright said that disk brake caliper remanufacturers take the used calipers, or cores, from distributors, refurbish them to a "same as new" condition, and then return them to distributors for sale to consumers. Mr. Wright said that this industry thus depends on the return of the cores from distributors and the ability to export remanufactured disk brake calipers to other countries. Mr. Wright said that in recent years, disk brake caliper remanufacturers have experienced more difficulty in obtaining credit because banks have been less willing to accept inventory as collateral for loans.

Finally, Mr. Wright stated that his company urges the following: (1) Congress should create a 20 percent tax credit for the purchase of equipment used to remanufacture products in the United States; (2) federal, state, and local governments should promote the use of remanufactured motor vehicle parts and components in their vehicle fleets; (3) USTR should continue to push for language to eliminate non-tariff barriers in all free trade agreements; and (4) interested parties should join together to create opportunities for public-private partnerships to conduct research and development to improve manufacturing technologies and practices to recycle and remanufacture vehicle components and parts.

APPENDIX E: MANUFACTURING ACTIVITIES QUESTIONNAIRE

Manufacturing and Remanufacturing Activities Questionnaire

Written Completion Version
United States International Trade Commission
Attention: REMAN Project Team
Office of Industries, Room 511
500 E Street, SW, Washington, DC 20436; Fax: 202-205-2217

The U.S. International Trade Commission (USITC) is conducting a fact-finding investigation to examine industries that produce or trade manufactured goods to find the extent of remanufacturing activities in the United States.

This investigation has been requested by the United States Trade Representative (USTR), who has directed the USITC to prepare a report that (1) provides an overview of the U.S. remanufactured goods industries and markets, (2) estimates U.S. and global trade in remanufactured goods, and (3) examines factors affecting trends in remanufactured goods trade.

According to the USTR, such data and analysis will help the United States pursue discussions and negotiations on eliminating barriers to trade in remanufactured goods and also be of benefit to the U.S. remanufacturing industry. This questionnaire has been designed to collect information to fulfill this request.

More information about this report and the investigation under which it is being prepared (No. 332-525) can be found on our web site at: *http://www.usitc.gov /research_and_analysis /What_We_Are_Working_On.htm*

Your firm is required by law to respond to this questionnaire

Please read all instructions and return the completed questionnaire to the USITC no later than March 2, 2012

The information is requested under the authority of section 332(g) of the Tariff Act of 1930 (19 U.S.C. § 1332(g)). Completing the questionnaire is mandatory, and failure to reply as directed can result in a subpoena or other order to compel the submission of records or information in your possession (19 U.S.C. § 1333(a)). Further information on this questionnaire can be obtained from the following project team members:

Alan Treat (202-708-4721; *alan.treat@usitc.gov*)
Jeremy Wise (202-708-2719; *jeremy.wise@usitc.gov*)

Confidentiality

The Commission has designated as "confidential business information" the information you provide in response to this questionnaire to the extent that such information would reveal the operations of your firm and is not otherwise available to the public. The Commission will not disclose such confidential business information except as provided for in section 10 of this questionnaire. Information received in response to this questionnaire will be aggregated with information from other questionnaire responses and will not be published in a manner that would reveal the operations of your firm.

Instructions

Completing the Questionnaire

1. **Retrieving the written completion version of the questionnaire.** If you need another copy of the questionnaire, please contact the project team (see cover for

contact information). This version of the questionnaire is appropriate if you are completing the questionnaire using written responses. An electronic completion version of this questionnaire is also available. It has been designed to simplify the entry process and minimize the need for our staff to contact your firm for clarifications. If your firm would prefer to use the electronic completion version, please go to the address below using a web browser and download it to your computer. *http://www.usitc.gov /documents/usitc-reman.pdf*

2. **Entering information.** Provide responses for each question that applies to your firm. Write in a response or check a box as indicated in each question.
3. **Numeric data.** Note that data for sales, employees, shipments, exports, and imports should be entered as full figures, not in thousands, millions, or similar format.
4. **Submitting the questionnaire.** After completing the questionnaire, follow the submission instructions in section 11. Please keep a copy of your submission for your records.

General

1. **Types of firms that should complete this questionnaire.** This questionnaire is intended for firms that have "manufactured goods activities"--in other words, firms that produce, sell, or trade manufactured goods. These include both firms that are headquartered in the United States and those that are U.S. affiliates of companies that are headquartered outside the United States. Firms that have manufacted goods activities, but no remanufactured goods activities, will only have to complete a few questions in sections 1, 2, and 3 before submitting their response.
2. **Coordinating your firm's response.** If separate persons or departments within your firm will share responsibility for completing this questionnaire, please make sure they coordinate their response so that the information your firm gives us is internally consistent. In our experience with past questionnaires, this will minimize our need to call you back for clarifications.
3. **Relationship to corporate structure.** Please provide a single response for your firm's activities. This may require your firm to combine information from two or more business units. If it is not possible to combine responses, or it is unreasonably burdensome, then your firm may provide separate responses for business units, but please ensure that there is no double counting. Joint venture firms should submit a separate response, and there should be no double counting with other business units of the joint venture partners.
4. **U.S. affiliates of foreign companies.** Please respond as if the affiliate were an independent firm operating in the United States. For example, show total sales for the affiliate only, and not for the foreign corporation. When listing investments, include both investments the affiliate has made in the United States, and investments in foreign countries that can be directly linked to the U.S. affiliate.
5. **Questionnaire structure.** This questionnaire is composed of 11 sections, as shown below.

Table of Contents

Definitions

1 Remanufactured goods:

Non-agricultural goods that are entirely or partially comprised of parts that (i) have been obtained from the disassembly of used goods; and (ii) have been processed, cleaned, inspected, and tested to the extent necessary to ensure they have been restored to original working condition or better; and for which the remanufacturer has issued a warranty. Some firms may use different terms for remanufactured, such as "refurbished," "rebuilt," "overhauled," etc. But if your firm's activities meet the definition above, then your firm has remanufactured goods activities for the purposes of this questionnaire, even if your firm or industry uses a different term of definition.

2. Cores:

Used goods that are the primary component input for remanufactured goods. Typically, these used goods are at the end of their useful life. Some industries may use different terms for cores, such as "casings" (the term used in tire retreading). But the definition applies, even if the term is not the same as used in your firm's industry.

3 Sectors and sector products:

Many of the questions in this questionnaire refer to industry sectors. These sectors, and product examples for each, are shown on the next page.

Question 3.7 requests that your firm indicate the sector(s) that reflect your remanufacturing activities. These sectors should be used in subsequent questions in sections 4, 6, and 7.

Other definitions appear throughout this questionnaire.

4 The sectors and sector descriptions used in this questionnaire are shown below. Top product examples for each sector are also shown.

Sector number	Sector	Sector description	Top product examples
1	Aerospace	Components, parts, modules, systems, or subsystems used in the production of airplanes and other aerospace equipment or their subsystems	Landing gears, flight surface actuators, engines, fuel systems, etc.
2	Motor vehicles	Components, parts, modules, systems, or subsystems used in the production of passenger cars and light-, medium- and heavy-duty on-road trucks (not including tires)	Starters, alternators, gear boxes, engines, differentials, brakes, etc.
3	Heavy-duty and construction equipment	Components, parts, modules, systems, or subsystems used in the production of construction, mining, and agriculture equipment	Diesel engines, transmissions, differential carriers, starters, brakes, etc.
4	Consumer electronics	Components, parts, modules, systems, or subsystems used in the production of consumer electronic products (not including desktop or laptop computers or tablets)	Mobile phones, cordless phones, smart phones, digital cameras, DVD players, televisions, etc.
5	Information technology (IT) equipment	Components, parts, modules, systems, or subsystems used in the collection, storage, and transmission of information, including telecommunications networks (not including mobile phones)	Desktop computers, laptops, servers, data storage arrays, network switches, routers, modems, base transceiver stations, etc.
6	Imaging products	Components, parts, modules, systems, or subsystems used in the production of imaging products	Copiers, printers, scanners, fax machines, ink and toner cartridges, fuser assemblies, etc.
7	Consumer household appliances	Components, parts, modules, systems or subsystems used in the production of consumer appliances	Washers, dryers, refrigerators, kitchen appliances, etc.
8	Machinery	Components, parts, modules, systems, or subsystems of industrial machinery used to produce other products, devices, or technologies	Industrial valves, turbines, machine tools, textiles machinery, compressors, etc.
9	Electrical apparatus	Components, parts, modules, systems, or subsystems used in the production of products associated with the regulation, distribution, and transmission of electrical energy	Power distribution conductors, transformers, switch gears and boards, etc.
10	Medical equipment	Components, parts, modules, systems, or subsystems of medical equipment used in the health care industry	Diagnostic, surgical, and patient care apparatus; scanners, medical pumps, x-ray equipment, etc.
11	Furniture	Components, parts, modules, systems, or subsystems used in the production of modular systems and stand-alone metal and wood furniture	Office systems furniture, laminate work surfaces, reupholstered office seating, etc.
12	Locomotives	Components, parts, modules, systems, or subsystems used in locomotives and rolling stock	Locomotive engines, drive motors, axles, other subsystems, etc.
13	Tires	Retreading of tires used in any industry	Tires for trucks, cars, off-the-road vehicles, airplanes, etc.
14	Restaurant equipment	Components, parts, modules, systems, or subsystems of equipment used in the food retail and food production industries	Ovens, beverage dispensers, food preparation tables, etc.
15	Other	Components, parts, modules, systems or subsystems used in equipment not elsewhere specified	

Section 1: Contact information

1.1. Please provide your firm's address and a contact person.

Firm name

Address

City State Zip code Web site address (www.name.domain)

Contact person's name Contact person's job title

Contact person's telephone number (xxx-xxx-xxxx) Contact person's e-mail address (xxx@xxx.xxx)

1.2. Does the firm named above have any joint ventures?

☐ Yes

☐ No

☐ Not applicable
(check this box if the firm named above is a joint venture)

If your firm has joint venture operations, each should provide a separate questionnaire response.

Section 2: Basic Information

The questions in this section refer to all of your firm's manufacturing activities in the United States, not just remanufacturing activities.

We need this information so we can compare and contrast the types of manufacturing activities involved.

Manufacturing definition: The mechanical, physical, or chemical transformation of materials, substances, or components into products, including related assembly operations. Manufacturing establishments are often described as plants, factories, or mills.

If your firm has just activities related to agriculture (e.g., crop growing), mining, oil/gas extraction, utilities (e.g., power generation), construction, or services, then your firm does not have manufacturing activities.

2.1. Did your firm produce or sell manufactured goods, or trade (import or export) new or used goods at any time during 2009-2011?

If so, select "Yes." Also select "Yes" if your firm has any activities related to cores used in remanufacturing, or remanufacturing wholesale or distribution activities.

☐ Yes. Continue to next question

☐ No. Skip to section 10

2.2. Please indicate the sector(s) in which your firm is active. Pick from the list of sectors shown on p. 5, if applicable. Otherwise write-in your own sector(s). If your firm participates in more than one sector, indicate these as well, up to a maximum of three sectors.

Primary sector	
Secondary sector	
Third sector	

Total sales definition: All sales, net of returns, discounts, and allowances. Includes internal consumption and transfers to related firms, as applicable, at fair market value. Same as sales as shown on a typical income statement.

Employee and full-time equivalent (FTE) definitions: Includes the number of your firm's employees at facilities located in the United States. Include production and related workers, managers, supervisors, technicians, office workers, etc. related to manufacturing activity on a full-time equivalent (FTE) basis.

If your firm is an affiliate of a foreign firm, include only employees that can be attributed directly to your firm's U.S. operations. Full-time equivalent (FTE) reflects the total number of regular straight-time hours (i.e., not including overtime or holiday hours) worked by employees divided by the number of compensable hours applicable to each calendar year. Annual leave, sick leave, and compensatory time off and other approved leave categories are considered to be “hours worked” for purposes of defining FTE employment.

2.3. Please list (1) the value of your firm's total sales for all activities and (2) the number of employees (on a full-time equivalent (FTE) basis) for your firm's operations **in the United States**. Careful estimates are acceptable.

If your firm is an affiliate of a foreign firm, include only sales and employees that can be directly attributed to your firm's U.S. operations. If a figure is zero, enter "0". If unknown, leave the entry box blank.

#	Item	2009	2010	2011
1	Total sales for all activities in the United States (in full figure dollars, not thousands or other format)			
2	Total number of employees engaged in activities in the United States, FTE basis (in full figures, not thousands or similar format)			

Section 3: Product Information

3.1. **For remanufactured goods only,** and the processes to make and handle these goods, please indicate below which activities your firm engaged in during 2009-2011 **in the United States**. Check all that apply. If your firm wholly contracted out an activity, do not check the box for that activity.

☐ Disassembled cores into parts

☐ Cleaned parts

☐ Inspected parts

☐ Tested parts

☐ Produced remanufactured goods

☐ Sold remanufactured goods

☐ Imported remanufactured goods

☐ Exported remanufactured goods

☐ Warehoused cores, remanufactured goods or parts

☐ Distributed cores, remanufactured goods or parts

☐ Imported cores

☐ Exported cores

☐ Purchased cores from a domestic firm

3.2. Did you check any of the boxes in the previous question?

☐ Yes. For the purposes of this questionnaire, your firm is considered to have remanufacturing activities. **Continue to the next question.**

☐ No. Skip to section 10

3.3. If you produce remanufactured goods, does the quality of these goods typically (check one): Equal the original working condition of the good Exceed the original working condition of the good

3.4. At anytime during 2009-2011 did your firm produce goods for which you did not provide a warranty but which otherwise met the definition of remanufactured goods?

☐ Yes

☐ No

3.5. **For remanufactured goods only,** and the processes to make and handle these goods, please indicate below which activities your firm engaged in during 2009-2011 in countries **other than the United States**. Check all that apply.

☐ Disassembled cores into parts

☐ Cleaned parts

☐ Inspected parts

☐ Tested parts

☐ Produced remanufactured goods primarily for the U.S. market

☐ Produced remanufactured goods primarily for markets other than the United States

☐ Sold remanufactured goods in a foreign country

☐ Imported remanufactured goods from countries other than the United States

☐ Exported remanufactured goods to countries other than the United States

☐ Warehoused cores, remanufactured goods or parts

☐ Distributed cores, remanufactured goods or parts

☐ Imported cores from countries other than the United States

☐ Exported cores to countries other than the United States

☐ Purchased cores from a supplier in a foreign country

3.6. A. Did one or more foreign entities have a substantial (10 percent or greater) ownership interest in your firm as of December 31, 2011? (An entity includes a corporation, limited liability company, partnership, sole proprietorship, or any subsidiary related to the foregoing.)

☐ Yes. Continue with part B of this question

☐ No. Go to next question

B. If you responded "Yes" to part A, indicate the country of origin of the foreign entities, up to a maximum of three. Top country Second country Third country

C. Does at least one of these foreign entities have remanufacturing activities in countries other than the United States?

☐ If Yes, continue with part D of this question

☐ No. Go to next question

D. If you responded "Yes" to part C, indicate the country in which these activities take place, up to a maximum of three.

Country 1	
Country 2	
Country 3	

E. Is the combined foreign ownership interest in your firm as of December 31, 2011 greater than 50 percent?

☐ Yes

☐ No

3.7. Please indicate the sector(s) in which your firm has **remanufacturing activities.** Pick from the list of sectors shown on p. 5, up to a maximum of three. Indicate the primary sector in part A. If your firm's activities are in other sectors as well, use parts B and, if needed, C below. Then indicate your firm's three most important remanufactured products in these sector(s). If your firm's sector(s) do not appear in the sectors shown on p. 5, write-in your own sector.

A. Primary sector	
Primary sector	
Product 1	
Product 2	
Product 3	

B. Second sector	
Second sector	
Product 1	
Product 2	
Product 3	

C. Third sector	
Product 1	
Product 2	
Product 3	

3.8. **Other than remanufacturing**, in which of these activities did your firm engage within the United States during 2009-2011? Check all that apply.

☐ Produced new goods in the United States

☐ Sold new goods in the United States

3.9. A. In any one year during 2009-2011 did your firm have a total of 20 or more employees in the United States engaged in **remanufactured goods activities** (including producing, selling, importing, exporting, warehousing, distribution, and related support activities)?

☐ Yes

☐ No

B. In any one year during 2009-2011 did your firm have a total of 5 or more employees in the United States engaged in activities related to the **import or export of remanufactured goods or cores**?

☐ Yes

☐ No

If you responded "Yes" to either part A or B in question 3.9 above, then proceed to the next section. If you responded "No" to both part A and B in question 3.9 above, then skip to section 9.

Section 4: Production, Shipments, and Employment

In this section, we are requesting these data to help us track trends in remanufacturing activities by sector. The sector(s) you list below should match the sector(s) you indicated in question 3.7 in the previous section. If you need to change the sector(s), please return to this question and make the appropriate change.

Report values in full figures, not in thousands or similar format.

Not all the questions in this section may apply to your firm. For example, if your firm has just import activities, then you would not respond to question 4.1. But it would be appropriate to respond to question 4.7 about your firm's import-related employment.

Shipments definition: Includes commercial shipments, internal consumption, and transfers to related firms. Shipment values should be reported as net values (i.e., gross sales values less all discounts, allowances, rebates, prepaid freight, and the value of returned goods), f.o.b. your U.S. point of shipment.

Tolling definition: Agreement between two firms whereby the first firm furnishes the core and the second firm uses the core to produce a remanufactured good that it then returns to the first firm with a charge for processing costs, overhead, etc.

4.1. Please list the value of your firm's total shipments of remanufactured goods to domestic and foreign markets broken down by sector. Include only goods **produced by your firm in facilities located in the United States**. Careful estimates are acceptable. If your firm is an affiliate of a foreign firm, include only production that can be directly attributed to your U.S. facilities. If a value is zero, enter "0". If unknown, leave the entry box blank.

Do not include core charges or deposits (if applicable).

Do not include shipments of remanufactured goods produced by other firms under a tolling or similar agreement. However, if your firm itself produced remanufactured goods under a tolling or similar agreement, then do report these shipments.

#	Sector (should match sector(s) from question 3.7)	2009	2010	2011
		Shipments from U.S. facilities (in full figure dollars)		
Shipments of remanufactured goods produced in the United States (do not include core charges or deposits) by sector:				
1				
2				
3				
4	All other remanufactured goods			
5	Total shipments of remanufactured goods produced in the United States			

4.2. During 2011, what share of your firm's U.S. shipments as reported in question 4.1 includes remanufactured goods made by other companies? (For example, if a particular shipment is an engine that contains an oil pump remanufactured by another company, what is the value of the pump as a share of the value of the engine?) Careful estimates are acceptable. Check one. If your firm did not have shipments of remanufactured goods produced in the United States in 2011, leave check boxes blank.

☐ None (or < 1 percent)

☐ 1-25 percent

☐ 26-50 percent

☐ 51-75 percent

☐ 76-100 percent

4.3. Based on the value of your firm's 2011 shipments reported in question 4.1, please list the top U.S. states (up to three) where the remanufactured goods were produced. If your firm did not have shipments of remanufactured goods produced in the United States in 2011, leave entry boxes blank.

State with the most production	
State with the second most production	
State with the third most production	

4.4. Please list the value of your firm's 2011 shipments reported in question 4.1, broken down according to the type of customers the remanufactured goods were sold to (OEM =

original equipment manufacturer). If your firm did not have shipments of remanufactured goods produced in the United States in 2011, leave entry boxes blank.

Customer		Shipments (in full figure dollars)
OEMs (including internal transfers)		
Non-OEM users (including internal transfers)		
Dealers (whether owned by your firm or independent)		
Wholesalers or retailers		
Other-Specify:		

4.5. Please indicate which of the following describes the changes in your firm's U.S. production capacity for remanufactured goods during 2009-2011. Check all that apply.

☐ No change

☐ Expanded production capacity at existing establishments

☐ Reduced production capacity at existing establishments

☐ Acquired production establishments

☐ Sold or otherwise disposed of production establishments

4.6. For your firm's 2011 U.S. shipments of remanufactured goods as reported in question 4.1, please indicate the share that were sold to the U.S. armed services (the U.S. Department of Defense including the Army, Navy, Air Force, Marines or reserve guard elements of the foregoing, U.S. Coast Guard, and the National Guard). Check one. If your firm did not have shipments of remanufactured goods produced in the United States in 2011, leave entry boxes blank.

None (or < 1 percent)

1-25 percent ☐

26-50 percent ☐

51-75 percent ☐

76-100 percent ☐

4.7. Please list the number of your firm's employees on a full-time equivalent (FTE) basis in either parts A or B as applicable. Careful estimates are acceptable. If your firm is an affiliate of a foreign firm, include only employees who can be directly attributed to your firm's U.S. activities. If a value is zero, enter "0". If unkown, leave the entry box blank. See page 7 for employee and full-time equivalent (FTE) definitions

A. Only firm's that reported shipments in question 4.1 should report employment in this part.

Provide the number of your firm's full-time equivalent (FTE) employees related to your firm's remanufacturing production by sector at facilities located in the United States during 2009-2011. Provide these figures by sector to the extent possible. If your firm has employees who can't be attributed solely to one sector or solely to remanufacturing, use the appropriate row below.

#	Sector (should match sector(s) from question 3.7)	2009	2010	2011
		Number of employees (FTE basis)		
Employees in remanufacturing production activities in the United States by sector:				
1				
2				
3				
4	All other sectors			
5	Employees who cannot be attributed by sector			
6	Employees who cannot be attributed solely to remanufacturing activities (please use this category only if absolutely necessary)			
7	Total			

B. Firms that have remanufacturing activities but no remanufacturing production should report employment in this part.

Provide the number of your firm's full-time equivalent (FTE) employees related to your firm's remanufacturing activities at facilities located in the United States during 2009-2011. If your firm has employees who can't be attributed solely to remanufacturing activities, use the appropriate row below.

#	Category	2009	2010	2011
		Number of employees (FTE basis)		
Employees offirms that do not produce remanufactured goods but have other remanufacturing activities:				
1	Employees who can be attributed solely to remanufacturing activities			
2	Employees who cannot be attributed solely to remanufacturing activities (please use this category only if absolutely necessary)			
3	Total			

Section 5: Investment

We are requesting the data in this section to help us track trends in remanufacturing investment. Note that you don't need to break down the data by sector. Please read the questions carefully to differentiate the investment types. In question 5.1, investment includes expenditures on property, plant, and equipment. In questions 5.2 and 5.3, investment includes expenditures related to ownership of another firm (also known as foreign direct investment).

Report values in full figures, not in thousands or similar format.

5.1. Please provide the value of any investments your firm made in remanufacturing activities made during 2009-2011 in the United States for the categories shown below. **Do not include capital expenditures used to acquire establishments.** Do not include investments, if any, made in your firm's foreign entities. Careful estimates are acceptable. If your firm is an affiliate of a foreign firm, include only capital expenditures that can be attributed directly to your U.S. operations. If a value is zero, enter "0". If unknown, leave the entry box blank.

This investment includes capital expenditures for plant construction, improvements to existing plant and equipment, and purchases of new or existing plant, property, machinery and equipment. Include direct expenditures by your firm, and expenditures by other firms (e.g., construction companies) done for your firm. Capital expenditures for R&D-related property, plant, and equipment are included in this category. Expenses for routine maintenance and repair are not included. Investments in financial instruments are also not included.

#	Category	2009	2010	2011
		Investment (in full figure dollars)		
1	Investment by your firm in domestic activities outside of remanufacturing			
2	Investment by your firm in domestic remanufacturing activities			
3	Unable to differentiate			
	Total investment in domestic activities			

5.2. Please identify the three largest foreign country sources of investment in your firm's U.S. remanufacturing activities and the value of such investments during 2009-2011. Careful estimates are acceptable. If your firm is an affiliate of a foreign firm, include only investments made in your U.S. operations. If a value is zero, enter "0". If unknown, leave entry box blank.

Include investment funds from a foreign entity used to purchase a 10 percent or greater ownership control in your firm and, if applicable, any subsequent funds from the foreign entity used to increase this ownership share. Also include any funds the foreign entity lent to your firm and earnings owed to the foreign entity that were reinvested in your firm.

#	Country	2009	2010	2011
		Investment (in full figure dollars)		
1				
2				
3				

4	All other countries			
5	Unable to differentiate			
	Total foreign investment			

5.3. Please identify the three largest foreign country destinations of your firm's foreign investment in remanufacturing activities and the value of such investments during 2009-2011. Careful estimates are acceptable. If your firm is an affiliate of a foreign firm, include only investments that can be attributed directly to your U.S. operations. If a value is zero, enter "0". If unknown, leave entry box blank.

#	Country	2009	2010	2011
		Investment (in full figure dollars)		
1				
2				
3				
4	All other countries			
5	Unable to differentiate			
	Total foreign investment			

Include investment funds your firm used to purchase a 10 percent or greater ownership control in a foreign entity and, if applicable, any subsequent funds your firm used to increase this ownership share. Also include any funds your firm lent to the foreign entity and earnings owed to your firm that were reinvested in the foreign entity.

Section 6: Remanufactured Goods Trade

We are requesting the data in this section to help us track trends in U.S. trade in remanufactured goods. Note that question 6.1 is composed of parts A-D on four separate pages so that you can break down the data by sector and by destination country. The sector(s) you indicate in parts A-D should match the sector(s) shown in question 3.7.

Report values in full figures, not in thousands or similar format.

6.1 Please provide the value of your firm's U.S. exports of remanufactured goods by sector and destination country during 2009-2011. Include exports made by your firm directly and exports of your firm's products by other firms. Careful estimates are acceptable. If a value is zero, enter "0". If unknown, leave the entry box blank.

Use the value as declared on export documentation.

A. For primary sector exports (indicate sector):

#	Country	2009	2010	2011
		Exports (in full figure dollars)		
Exports to U.S. free trade partners:				
1	Australia			
2	Bahrain			
#	Country	2009	2010	2011
		Exports (in full figure dollars)		
3	Chile			
4	Canada			
5	DR-CAFTA: Dominican Republic, Costa Rica, El Salvador, Guatemala, Honduras, Nicaragua			
6	Israel			
7	Jordan			
8	Mexico			
9	Morocco			
10	Oman			
11	Peru			
12	Singapore			
Exports to other countries (indicate top countries, up to a maximum of 3):				
13				
14				
15				
16	All other countries			
If your firm is unable to differentiate exports by country:				
17	Unable to differentiate			
18	Total exports			

B. For second sector exports (indicate sector):

#	Country	2009	2010	2011
		Exports (in full figure dollars)		
Exports to U.S. free trade partners:				
1	Australia			
2	Bahrain			
3	Chile			
4	Canada			

5	DR-CAFTA: Dominican Republic, Costa Rica, El Salvador, Guatemala, Honduras, Nicaragua			
6	Israel			
7	Jordan			
8	Mexico			
#	Country	2009	2010	2011
		Exports (in full figure dollars)		
9	Morocco			
10	Oman			
11	Peru			
12	Singapore			
Exports to other countries (indicate top countries, up to a maximum of 3):				
13				
14				
15				
16	All other countries			
If your firm is unable to differentiate exports by country:				
17	Unable to differentiate			
18	Total exports			

C. For third sector exports (indicate sector):

#	Country	2009	2010	2011
		Exports (in full figure dollars)		
Exports to U.S. free trade partners:				
1	Australia			
2	Bahrain			
3	Chile			
4	Canada			
5	DR-CAFTA: Dominican Republic, Costa Rica, El Salvador, Guatemala, Honduras, Nicaragua			
6	Israel			
7	Jordan			
8	Mexico			
9	Morocco			

10	Oman			
11	Peru			
12	Singapore			
Exports to other countries (indicate top countries, up to a maximum of 3):				
13				
14				
15				
16	All other countries			
		Exports (in full figure dollars)		
If your firm is unable to differentiate exports by country:				
17	Unable to differentiate			
18	Total exports			

D. For all other sectors

#	Country	2009	2010	2011
		Exports (in full figure dollars)		
Exports to U.S. free trade partners:				
1	Australia			
2	Bahrain			
3	Chile			
4	Canada			
5	DR-CAFTA: Dominican Republic, Costa Rica, El Salvador, Guatemala, Honduras, Nicaragua			
6	Israel			
7	Jordan			
8	Mexico			
9	Morocco			
10	Oman			
11	Peru			
12	Singapore			
Exports to other countries (indicate top countries, up to a maximum of 3):				
13				
14				
15				
16	All other countries			

If your firm is unable to differentiate exports by country:				
17	Unable to differentiate			
18	Total exports			

6.2. Please provide the value of your firm's U.S. imports of remanufactured goods by sector. Include imports made by your firm directly and imports made on your firm's behalf by other firms. Careful estimates are acceptable. If a value is zero, enter "0". If unknown, leave the entry box blank.

Use the value as declared on import documentation.

#	Sector (should match sector(s) from question 3.7)	2009	2010	2011
		Imports (in full figure dollars)		
Remanufactured goods imports:				
1				
2				
3				
4	All other sectors			
5	Unable to differentiate by sector			
6	Total imports of remanufactured goods			

6.3. For the remanufactured goods imports you reported in question 6.2, what countries were your firm's leading import suppliers in 2011? Indicate the countries below, up to a maximum of three. The sum of the imports reported in this question should not exceed total imports for 2011 as shown on line 6 of question 6.2 above.

Rank	Country	Imports (in full figure dollars)
Top country		
Second leading country		
Third leading country		

Section 7: Cores

Report values in full figures, not in thousands or similar format.

7.1. Please provide the value of the cores purchased by your firm from domestic sources that were used in the production of remanufactured goods in facilities located in the United States during 2009-2011 by sector. Careful estimates are acceptable. If a value is zero, enter "0". If unknown, leave the entry box blank. Include, for example, cores purchased from a core dealer. But do not include cores acquired from customers or other firms that did not require a payment other than a deposit.

#	Sector (should match sector(s) from question 3.7)	2009	2010	2011
		Cores purchased (full figure dollars)		
Cores purchased from domestic sources and used to produce remanufactured goods:				
1				
2				
3				
4	All other sectors			
5	Total value of cores purchased			

7.2. Please indicate the leading sources (up to a maximum of three) of the cores your firm acquired in 2011 by ranking the following from 1 to 3 (1 = largest source). For the purpose of this question, do not include small sources (less than 5 percent of the number of cores supplied) in your ranking.

Core source	Rank top 3 (select 1, 2, and 3 as applicable)
Warranty returns	
Core dealers	
Trade-in returns from customers or dealers	
Returned by owner for remanufacturing	
Salvage operators, dismantlers, recyclers	
Imports	
Other-specify:	

7.3. Please provide the value of your firm's U.S. exports to foreign countries of cores to be used in remanufactured goods during 2009-2011 by sector. Include exports made by your firm directly and exports by other firms of your firm's cores. Careful estimates are acceptable. If a value is zero, enter '0". If unknown, leave the entry box blank.

Use the value as declared on export documentation.

#	Sector (should match sector(s) from question 3.7)	2009	2010	2011
		Core export value (in full figure dollars)		
Core exports:				
1				
2				
3				
4	All other sectors			
5	Total exports of cores			

7.4 For the exports of cores that you reported in question 7.3, which countries were your firm's leading export destinations in 2011? Indicate the countries below, up to a maximum of three, and indicate the value of these exports. The sum of the exports reported in this question should not exceed total exports for 2011 as shown on line 5 of question 7.3 above.

Rank	Country	Exports (in full figure dollars)
Top country		
Second leading country		
Third leading country		

7.5 Please provide the value of your firm's U.S. imports from foreign countries of cores to be used in remanufactured goods during 2009-2011 by sector. Include imports made by your firm directly and imports made on your firm's behalf by other firms. Careful estimates are acceptable. If a value is zero, enter "0". If unknown, leave the entry box blank.

Use the value as declared on import documentation

#	Sector (should match sector(s) from question 3.7)	2009	2010	2011
		Core import value (in full figure dollars)		
Core imports:				
1				
2				
3				
4	All other sectors			
5	Total imports of cores			

7.6 For the imports of cores that you reported in question 7.5, which countries were your firm's leading import suppliers in 2011? Indicate the countries below, up to a maximum of three, and indicate the value of these imports. The sum of the imports reported in this question should not exceed total imports for 2011 as shown on line 5 of question 7.5 above.

Rank	Country	Imports (in full figure dollars)
Top country		
Second leading country		
Third leading country		

Section 8: Market information

8.1 Indicate how important the following factors are in affecting your firm's ability to compete in the U.S. market for remanufactured goods. If your firm has no knowledge about the factor, leave all boxes for that factor unchecked.

#	Factor	Rank significance from 1 (not at all important) to 5 (extremely important)				
		Not at all important	Somewhat important	Important	Very important	Extremely important
1	High price of cores	1 □	2 □	3 □	4 □	5 □
2	Scarcity of cores	1 □	2 □	3 □	4 □	5 □
3	Labor costs	1 □	2 □	3 □	4 □	5 □
4	Scarcity of skilled workers	1 □	2 □	3 □	4 □	5 □
5	Difficulty accessing capital	1 □	2 □	3 □	4 □	5 □
6	Borrowing costs	1 □	2 □	3 □	4 □	5 □
7	Energy costs	1 □	2 □	3 □	4 □	5 □
8	Research and development costs	1 □	2 □	3 □	4 □	5 □
9	Transportation costs	1 □	2 □	3 □	4 □	5 □
10	Health care costs	1 □	2 □	3 □	4 □	5 □
11	Unfavorable tax treatment	1 □	2 □	3 □	4 □	5 □
12	Environmental regulations	1 □	2 □	3 □	4 □	5 □
13	Customer preference for new products	1 □	2 □	3 □	4 □	5 □
14	Declining demand for remanufactured goods	1 □	2 □	3 □	4 □	5 □
15	Competition from imported remanufactured goods	1 □	2 □	3 □	4 □	5 □
16	Availability of low cost new products	1 □	2 □	3 □	4 □	5 □
17	Licensing or certification requirements	1 □	2 □	3 □	4 □	5 □
18	Other-specify:	1 □	2 □	3 □	4 □	5 □
19	Other-specify:	1 □	2 □	3 □	4 □	5 □

8.2 Indicate how important the following factors are in affecting your firm's ability to compete in markets **other than the U.S. market**. If your firm has no knowledge of the factor, leave all boxes for that impediment unchecked.

#	Factor		Rank significance from 1 (not at all important) to 5 (extremely important)				
			Not at all important	Somewhat important	Important	Very important	Extremely important
1	High price of cores		1 □	2 □	3 □	4 □	5 □
2	Regulatory barriers in foreign market		1 □	2 □	3 □	4 □	5 □
3	Lack of knowledge of foreign market		1 □	2 □	3 □	4 □	5 □
4	Foreign market tariffs		1 □	2 □	3 □	4 □	5 □
5	Transportation costs		1 □	2 □	3 □	4 □	5 □
6	Lack of distribution and marketing channels		1 □	2 □	3 □	4 □	5 □
7	Cost of compliance with product environmental standards		1 □	2 □	3 □	4 □	5 □
8	Cost of compliance with other product standards		1 □	2 □	3 □	4 □	5 □
9	Licensing or certification requirements		1 □	2 □	3 □	4 □	5 □
10	Other-specify:		1 □	2 □	3 □	4 □	5 □

8.3. Indicate the foreign countries whose **regulatory** barriers pose the biggest problems in your firm's experience. Indicate a country in the first column (up to a maximum of three), then describe the country's must problematic barrier(s) to the right.

Foreign country	Describe barrier(s)

Section 9. Other information

9.1. If your firm would like to give more details about the competitiveness in the remanufacturing goods industry, including access to foreign markets, use the space below.

9.2. If your firm would like to give more details about any of the responses in this questionnaire, or provide any other pertinent information, use the space below.

Section 10. Certification

The undersigned certifies that the information supplied herein in response to this questionnaire is complete and correct to the best of his/her knowledge and belief and understands that the information submitted is subject to audit and verification by the USITC. Section 332(g) of the Tariff Act of 1930 (19 U.S.C. 1332(g)) provides that the Commission may not release information which it considers to be confidential business information unless the party submitting such information had notice, at the time of submission, that such information would be released by the Commission, or such party subsequently consents to the release of the information. The undersigned acknowledges that information submitted in this questionnaire response and throughout this investigation may be used by the USITC, its employees, and contract personnel who are acting in the capacity of USITC employees, for the purposes of developing or maintaining the records of this investigation or related proceedings for which this information is submitted, or in internal audits and in investigations relating to the programs and operations of the USITC pursuant to 5 U.S.C. Appendix 3. The undersigned understands that all contract personnel will sign nondisclosure agreements. The information your firm provides in response to this questionnaire will be treated by the Commission as confidential and will not be disclosed to the public unless required by law. The information will be aggregated with information from other questionnaire responses and will be published in a manner that will not reveal the operations of your firm. The USTR has asked that the Commission not include any confidential business information in the report it transmits to him.

Certifier's name and title

Date of certification

Certifier's signature

Section 11. Questionnaire submission

11.1. Before submitting your firm's completed questionnaire, please report the actual number of hours required and the cost to your firm of completing this questionnaire, including all preparatory activities.

Hours	Cost ($)

11.2. Please make sure that the following items have been addressed before submitting your firm's questionnaire.

All responders:

□ Section 1.1 indicates a company name and a contact person.

□ Responders that selected "Yes" for question 2.1:

A primary sector has been selected in question 2.2.

□ All value figures provided in question 2.3 are in full figures, not in thousands or similar format.

Responders that selected "Yes" for question 3.2:

□ A primary sector for remanufactured goods has been selected in question 3.7.

Responders that selected "Yes" to either part A or part B in question 3.9:

□ All value figures provided in sections 4-7 are in full figures, not in thousands or similar format.

11.3. Mail or fax the completed questionnaire to us (see address and fax number below). Sending the questionnaire by U.S. mail is not recommended because this type of mail undergoes additional processing to screen for hazardous material that will likely substantially delay the delivery. Overnight mail service is recommended.

United States International Trade Commission
Attention: REMAN Project Team
Office of Industries, Room 511
500 E Street, SW,
Washington, DC 20436;
Fax: 202-205-2217

APPENDIX F: DESCRIPTION OF USITC QUESTIONNAIRE METHODOLOGY

Description of USITC Sector Selection and Questionnaire Methodology

In his letter to the Chairman of the U.S. International Trade Commission (Commission), the United States Trade Representative (USTR) requested that the Commission's report focus on remanufacturing-intensive sectors that account for the majority of remanufacturing activity in the United States.

The USTR also requested that the Commission's report be based on a review of available data and other information, including the collection of primary data through a survey of enterprises engaged in remanufacturing. The Commission field-tested its questionnaire with firms in 2011 and submitted it to the Office of Management and Budget (OMB) for clearance in February 2012. After receiving OMB clearance in February 2012, the Commission sent the questionnaire to a sample of 7,000 firms.

Sector Selection Methodology

The USTR did not define "remanufacturing-intensive"; however, based on interviews with industry representatives and on a review of publicly available information, the Commission identified 12 industry sectors that are considered to be remanufacturing intensive (i.e., likely to do more remanufacturing than other industry sectors),[1] and that are estimated to account for the majority of remanufacturing activity in the United States.[2] In addition, Commission staff identified and included certain wholesaling activities in an effort to more accurately estimate trade in remanufactured goods by nonremanufacturing firms. These industry and wholesaling sectors were then mapped to North American Industry Classification System (NAICS) codes, as show in table F.1.

Table F.1. Industry sector mapping

Industry sector	2007 NAICS Code	Description
Aerospace	336411	Aircraft manufacturing
	336412	Aircraft engine and engine parts manufacturing
	336413	Other aircraft parts and auxiliary equipment manufacturing
	488190	Other support activities for air transportation
Motor vehicle parts	333618	Other engine equipment manufacturing
	3361	Motor vehicle manufacturing
	3363	Motor vehicle parts manufacturing
Heavy duty and off-road (HDOR) equipment	3331, except 333112[a]	Agriculture, construction, and mining machinery manufacturing

Industry sector		2007 NAICS Code	Description
		333995	Fluid power cylinder and actuator manufacturing
Consumer products	Consumer electronics	3342, except 334290[b]	Communications equipment manufacturing
		3343	Audio and video equipment manufacturing
	Consumer goods	339920	Sporting and athletic goods manufacturing
		339932	Game, toy, and children's vehicle manufacturing
	Consumer household appliances	3352	Household appliance manufacturing
Information technology (IT) products	IT equipment	333313	Office machinery manufacturing
		3341	Computer and peripheral equipment manufacturing
		334290	Other communications equipment manufacturing
		3344	Semiconductor and other electronic component manufacturing
		811213	Communication equipment repair
	Imaging products	325910	Printing ink manufacturing
		325992	Photographic film and chemical manufacturing
		333293	Printing machinery and equipment manufacturing
		333315	Photographic and photocopying equipment manufacturing
		811212	Computer and office machine repair
Machinery		332911	Industrial valve manufacturing
		332991	Ball and roller bearing manufacturing
		3332, except 333293[b]	Industrial machinery manufacturing
		333311	Automatic vending machine manufacturing
		333312	Commercial laundry and drycleaning manufacturing
		3334, except 333415[c]	HVAC and commercial refrigeration equipment
		333512	Metal cutting machine tool manufacturing
		333513	Metal forming machine tool manufacturing
		333516	Rolling mill machinery and equipment manufacturing

Table F.1. (Continued)

Industry sector	2007 NAICS Code	Description
	333518	Other metalworking machinery manufacturing
	3336, except 333618[d]	Turbine and power transmission equipment manufacturing
	3339, except 333995[e]	Other general purpose machinery manufacturing
Electrical apparatus	3353	Electrical equipment manufacturing
Medical devices	325413	In-vitro diagnostic substance manufacturing
	334510	Electromedical apparatus manufacturing
	334517	Irradiation apparatus manufacturing
	339112	Surgical and medical instrument manufacturing
	339113	Surgical appliance and supplies manufacturing
	339114	Dental equipment and supplies manufacturing
	339115	Ophthalmic goods manufacturing
Office furniture	3372, except 337212[f]	Office furniture and fixtures manufacturing
Locomotive systems	336510	Railroad rolling stock manufacturing
Tires	326211	Tire manufacturing, except retreading
	326212	Tire retreading
Restaurant equipment	333319	Other commercial and service machinery manufacturing
	333415	AC, refrigeration, and forced air heating
Wholesaling	423130	Tire and tube merchant wholesalers
	423140	Used motor vehicle parts merchant wholesalers
	423430	Computer and software merchant wholesalers
	423620	Electric appliance merchant wholesalers
	423690	Other electronic parts merchant wholesalers
	423930	Recyclable material merchant wholesalers

Source: USITC staff; USDOC, Census, NAICS, 2007.

[a]Excluded from the HDOR equipment sector is lawn and garden tractor and home lawn and garden equipment manufacturing.

[b]Classified in the IT products sector.

[c]Classified in the restaurant equipment sector.

[d]Classified in the motor vehicle parts sector.

[e]Classified in the HDOR equipment sector.

[f]Excluded from the furniture sector are customs architectural woodwork and millwork manufacturing.

Constructing the Sampling Frame

The sampling frame includes all manufacturing firms and select wholesaling firms that conduct business in the United States and that are classified under the NAICS codes for the industry sectors Commission staff identified as remanufacturing intensive.[3] It was compiled from the following sources: (1) Commission staff research to identify potential remanufacturers; (2) a database of firms likely engaged in remanufacturing activities obtained by Commission staff from the Rochester Institute of Technology (RIT), Rochester, NY; (3) a keyword search of likely remanufacturers in ORBIS, a commercial database that compiles statistical information on all known U.S. firms; (4) a keyword search of likely importers of remanufactured goods in PIERS, a commercial database containing information on oceangoing import and export trade; (5) all firms with 20 or more employees listed in ORBIS that are classified by primary NAICS codes belonging to industry sectors that Commission staff defined as remanufacturing intensive; (6) wholesaler firms in NAICS wholesaler categories that are considered likely to deal in remanufactured goods, including consumer electronics, information and technology (IT) products, motor vehicle parts, and retreaded tires. Based on this methodology, the estimated population size was 33,640 firms.[4]

Stratification and Sampling

The sample population of firms was stratified (divided into different subgroups of firms). The three main subgroups sorted firms by (1) information sources, (2) industry sector, and (3) size. The information source subgroup included firms that were identified either by Commission staff, by RIT, or by a keyword search in the ORBIS and PIERS databases. The industry sector subgroup included firms identified by NAICS codes in ORBIS. Firms identified by NAICS codes were further divided by size—small and medium-sized enterprises, or SMEs (20–499 employees) and large firms (500 or more employees). This yielded a total of 34 subgroups (or strata). Within each of these three main subgroups (by information source, industry sector, and firm size), the Commission employed a stratified random sample without replacement:

- Firms identified by information source were all sampled because Commission staff believed that these firms were extremely likely to engage in remanufacturing activities.
- Firms identified by industry sector were thought to be only somewhat likely to engage in remanufacturing activities. Large firms were all sampled, since there were relatively fewer of them, but SMEs were sampled randomly.[5]
- The resulting sample size was 7,000 firms. In choosing this sample size, the Commission sought to balances its desire for robust results with the burden placed on firms answering the survey.

Response Rate

Based on the Commission's legal authority under section 333(a) of the Tariff Act of 1930 (19 U.S.C. §1333(a)), the questionnaire was mandatory for all sampled firms to complete.

Besides the initial mailing, firms included in the sample received two follow-up mailings reminding them to complete the questionnaire. The Commission received 3,621 completed and timely questionnaires. In addition, the Commission received 205 responses by firms that indicated that the questionnaire did not apply to them, and 471 questionnaires were returned as undeliverable by the U.S. Postal Service.[6] Adjustments to the response rate were also made to account for duplicate responses by firms. In their questionnaire responses, some firms reported figures on a consolidated basis (i.e., including responses on behalf of their subsidiaries or affiliates), while others did not. In addition, some firms were identified more than once within the sample. As a result, there were 159 duplicates, primarily comprising multiple subsidiaries of the same firm. For firms sampled at 100 percent, duplicates were removed from the population values. For firms that were randomly sampled, the percent of duplicates (as a share of each stratum's population) was calculated and extrapolated to the entire stratum's population. The calculated share was then subtracted from the overall population. Commission staff conducted econometric tests to determine whether there was any observable non-response bias. Logistic regressions were run on a variety of data available for both respondents and non-respondents, including number of employees, revenues, and locate-on by state. No variable or sets of variables had significant explanatory power over firms' response or non-response.

Analysis of Questionnaire Responses

Once the Commission received completed questionnaires, they were reviewed by Commission staff to ensure that firms had properly reported all requested data. In cases where data were missing or appeared to be incorrect, staff contacted respondent firms to obtain corrected data.

Weighting

Due to sampling design, the raw data collected from completed questionnaires were not self-weighted. Rather, sampling weights were constructed to make the sample more closely reflect the population. Weights are used to account for firms that were either not sampled or did not respond to the questionnaire. Questionnaire responses were multiplied by a certain weight to project or extrapolate what a response might be for the entire population. Response rates that were high relative to the number of firms in a given stratum were assigned relatively lower weights and vice versa.[7] The weighting was based on strata and adjusted for duplicates (as described above):

$$w_s = \frac{N_s}{n_s}$$

where N_s is the adjusted population of stratum s and n_s is the number of respondents in stratum s. The weights ranged from 1.3 to 28.2 across strata.

Relative Standard Errors (RSE)

The set of firms selected for this survey was only one of many possible samples that could have been selected using the same sample methodology, and each of these possible samples would likely yield different results. In order to quantify the effects of sample-to-sample variation, relative standard errors (RSEs) were used to indicate the degree of precision conferred on a particular estimated value.[8] RSEs are defined as the standard error of a particular estimate divided by the estimate itself. A smaller RSE yields a more precise estimate. For instance, an RSE of 50 percent means that the standard error is half the size of the estimated value, whereas an RSE of 100 percent means that the standard error is equal to the size of the estimated value. Unless otherwise noted, all estimates based on weighted responses to the Commission's survey have an RSE below 55 percent. Estimates that produced higher RSEs (i.e., estimates that have less precision) are noted throughout the report. In the case where only a single observation was available for a particular stratum, a standard error could not be calculated. Instead, the single observation was assumed to have the same standard error as the average of the other observations in the other strata.[9]

REFERENCE

U.S. Department of Commerce (USDOC). U.S. Bureau of the Census (Census). "How the Data Are Collected." *Annual Survey of Manufacturers.* http://www.census.gov /manufacturing/asm/how_the_ data_are_collected /index.html (accessed August 22, 2012).

APPENDIX G: ADDITIONAL DATA TABLES

This appendix contains additional data tables on U.S. remanufacturing that were omitted from the chapters of the report, but that may still be of interest to the reader. Based on weighted responses to the Commission questionnaire, USITC staff estimated the following for the period 2009–11, broken down by firm size: (1) U.S. production of remanufactured goods, (2) U.S. remanufacturing investment, (3) U.S. remanufacturing employment, (4) U.S. exports and imports of remanufactured goods, (5) U.S. exports and imports of cores, and (6) U.S. inbound and outbound foreign direct investment (FDI). These data are shown in tables G.1–G.9. In addition, as noted in chapter 2, many U.S. remanufacturers use remanufactured goods as inputs in the production of other remanufactured products; as a result, some goods may be double-counted—once as an input, once as a finished good—in calculating production statistics. USITC staff estimated the percentage of firms that use remanufactured goods as inputs, as well as the 2011 share of the value of those inputs, in order to gauge how far double-counting may have overstated the 2011 value of production shipments, as reported by questionnaire respondents. These data are shown in tables G.10 and G.11. Finally, factors cited by U.S. remanufacturers as "extremely important" in affecting their ability to compete in foreign markets, by remanufacturing sector, are shown in table G.12.

Table G.1. Remanufactured goods: U.S. production of remanufactured goods by firm size, 2009–11

Firm size (by number of employees)	2009	2010	2011	% change, 2009–11
		Thousand $		
20–99	4,754,672	5,187,780	6,253,303	31.52
100–499	4,098,463	4,087,178	4,888,582	19.28
500+	28,180,232	30,383,196	31,607,215	12.16
Other[a]	247,417	223,830	251,046	1.47
Total production	37,280,784	39,882,004	43,000,147	15.34

Source: USITC staff calculations of weighted responses to the Commission questionnaire. Note: Totals may not sum due to rounding.

[a]Includes questionnaire respondents that either reported having fewer than 20 employees, or did not report their employment levels.

Table G.2. Remanufactured goods: U.S. remanufacturing investment by firm size, 2009–11

Firm size (by number of employees)	2009	2010	2011	% change, 2009–11
		Thousand $		
20–99	135,970	99,845	162,537	19.5
100–499	134,399	74,502	85,615	–36.3
500–up	440,898	845,528	967,604	119.5
Other[a]	2,460	3,472	7,570	207.7
Total investment	713,727	1,023,346	1,223,326	71.4

Source: USITC staff calculations of weighted responses to the Commission questionnaire. Note: Totals may not sum due to rounding.

[a]Includes questionnaire respondents that either reported having fewer than 20 employees, or did not report their employment levels.

Table G.3. Remanufactured goods: U.S. remanufacturing employment by firm size, 2009–11

Firm size (by number of employees)	2009	2010	2011	% change, 2009–11
	Full-time workers			
20–99	38,049	39,294	41,994	10.4
100–499	21,469	21,998	23,496	9.4
500+	103,067	107,967	111,072	7.8
Other[a]	3,245	2,883	2,946	–9.2
Total employment	165,830	172,142	179,508	8.2

Source: USITC staff calculations of weighted responses to the Commission questionnaire. Note: Totals may not sum due to rounding.

[a]Includes questionnaire respondents that either reported having fewer than 20 employees, or did not report their employment levels.

Table G.4. Remanufactured goods: U.S. exports of remanufactured goods by firm size, 2009–11

Firm size (by number of employees)	2009	2010	2011	% change, 2009–11
		Thousand $		
20–99	1,352,954	1,725,430	2,046,855	51.3
100–499	1,012,443	1,184,138	1,391,991	37.5
500+	5,038,774	5,783,355	8,166,921	62.1
Other[a]	102,703	98,244	114,402	11.4
Total	7,506,875	8,791,167	11,720,170	56.1

Source: USITC staff calculations of weighted responses to the Commission questionnaire.

Note: Data for U.S. exports of retreaded tires by firm size are not available. Total U.S. exports of remanufactured goods shown in this table may therefore differ from total U.S. imports of remanufactured goods that include retreaded tires as shown in chapter 2. Totals may not sum due to rounding.

[a]Includes questionnaire respondents that either reported having fewer than 20 employees, or did not report their employment levels.

Table G.5. Remanufactured goods: U.S. imports of remanufactured goods by firm size, 2009–11

Firm size (by number of employees)	2009	2010	2011	% change, 2009–11
		Thousand $		
20–99	566,698	726,698	885,796	56.3
100–499	745,926	1,047,226	898,266	20.4
500+	4,615,196	7,364,830	8,324,015	80.4
Other[a]	322,207	360,925	143,685	-55.4
Total	6,250,028	9,499,679	10,251,762	64.0

Source: USITC staff calculations of weighted responses to the Commission questionnaire.

Note: Data for U.S. imports of retreaded tires by firm size are not available.

Total U.S. imports of remanufactured goods shown in this table may therefore differ from total U.S. imports of remanufactured goods that include retreaded tires as shown in chapter 2.

Totals may not sum due to rounding.

[a]Includes questionnaire respondents that either reported having fewer than 20 employees, or did not report their employment levels.

Table G.6. Remanufactured goods: U.S. core exports by firm size, 2009–11

Firm size (by number of employees)	2009	2010	2011	% change, 2009–11
		Thousand $		
20–99	104,382	156,869	174,106	66.8
100–499	368,741	366,814	479,773	30.1
500+	1,003,310	1,245,296	138,5311	38.1
Other[a]	4,774	5,376	4,747	-0.6
Total	1,481,207	1,774,355	2,043,937	38.0

Source: USITC staff calculations of weighted responses to the Commission questionnaire.

Note: Data for U.S. exports of cores of retreaded tires (tire casings) by firm size are not available. Total U.S. core exports shown in this table may therefore differ from total U.S. core exports that include tire casings as shown in chapter 2.

Totals may not sum due to rounding.

[a]Includes questionnaire respondents that either reported having fewer than 20 employees, or did not report their employment levels.

Table G.7. Remanufactured goods: U.S. core imports by firm size, 2009–11

Firm size (by number of employees)	2009	2010	2011	% change, 2009–11
		Thousand $		
20–99	5,083	10,602	114,412	2,150.9
100–499	16,367	17,804	33,030	101.8
500+	752,338	1,000,790	1,546,550	105.6
Other[a]	322	796	1,587	392.9
Total	774,110	1,029,992	1,695,579	119.0

Source: USITC staff calculations of weighted responses to the Commission questionnaire.

Note: Data for U.S. imports of cores of retreaded tires (tire casings) by firm size are not available. Total U.S. core imports shown in this table may therefore differ from total U.S. core exports that include tire casings as shown in chapter 2. Totals may not sum due to rounding

[a]Includes questionnaire respondents that either reported having fewer than 20 employees, or did not report their employment levels.

Table G.8. Remanufactured goods: FDI in U.S. remanufacturing by firm size, 2009–11

Firm size (by number of employees)	2009	2010	2011	% change, 2009–11
		Thousand %		
20–99	901	176	2,489	176.2
100–499	193,647	196,575	229	–99.9
500+	82,396	481,471	453,502	450.4
Other[a]	246	1,061	1,768	618.7
Total	277,190	679,283	457,988	65.2

Source: USITC staff calculations of weighted responses to the Commission questionnaire. Note: Totals may not sum due to rounding.

[a]Includes questionnaire respondents that either reported having fewer than 20 employees, or did not report their employment levels.

Table G.9. Remanufactured goods: U.S. outbound FDI in foreign remanufacturing by firm size, 2009–11

Firm size (by number of employees)	2009	2010	2011	% change, 2009–11
		Thousand $		
20–99	8,483	5,187	8,556	0.9
100–499	47,887	36,596	29,854	–37.7
500+	61,013	181,578	244,750	301.1
Other[a]	1,836	2,552	4,005	118.1
Total	119,219	225,914	287,166	140.9

Source: USITC staff calculations of weighted responses to the Commission questionnaire. Note: Totals may not sum due to rounding.

[a]Includes questionnaire respondents that either reported having fewer than 20 employees, or did not report their employment levels.

Table G.10. Remanufactured goods: Percentage of U.S. remanufacturers that use remanufactured goods as inputs by sector

Sector	Percent
Aerospace	74.4
HDOR equipment	7.31
Medical devices	60.2
IT products	51.3
Machinery	50.5
Sector	Percent
Motor vehicle parts	37.8
Other[a]	30.9
Retreaded tires	2.1
Consumer products	([b])

Source: USITC staff calculations of weighted responses to the Commission questionnaire.

[a]Includes the electrical apparatus, locomotive, office furniture, and restaurant equipment remanufacturing sectors.

[b]None reported.

Table G.11. Remanufactured goods: Share of value contributed by remanufactured intermediate inputs by sector

	Share of value	
Sector	(%)	(thousands $)
Medical devices	22.4	328,436
Aerospace	18.4	2,397,788
HDOR equipment	17.7	1,372,715
Motor vehicle parts	16.0	994,097
Machinery	14.1	816,437
Other[a]	12.6	500,394
IT imaging equipment	11.4	305,317
Retreaded tires	0.7	10
Consumer products	([b])	([b])
Total	15.6	6,725,247

Source: USITC staff calculations of weighted responses to the Commission questionnaire.

[a]Includes the electrical apparatus, locomotive, office furniture, and restaurant equipment remanufacturing sectors. [b]None reported.

Table G.12. Remanufactured goods: Factors cited by U.S. remanufacturers that export as "extremely important" that affect their ability to compete in foreign markets by sector

Sector	Factor	Percent
IT products	Transportation costs	27.4
Medical devices	Transportation costs	24.6
Motor vehicle parts	High price of cores	15.2
HDOR equipment	Transportation costs	15.2
Consumer products	High price of cores	13.6
Aerospace	Transportation costs	13.5
Wholesale	Regulatory barriers in foreign markets	8.0
Machinery	Licensing or certification requirements	8.0
Retreaded tires	Transportation costs	2.1

Source: USITC staff calculations of weighted responses to the Commission questionnaire.

End Note for Executive Summary

[1] Because of differences in terminology across industry sectors, ambiguity in the definition of remanufactured goods, and the nonsystematic survey of importers and firms with fewer than 20 employees, this report's estimates of remanufactured goods production, trade, and employment may be somewhat low.

End Notes for Introduction

[1] See appendixes A and B, respectively, for the request letter from the USTR and the *Federal Register* notice associated with this investigation.

[2] Case studies include (in chapter order) the aerospace, heavy-duty and off-road (HDOR) equipment, motor vehicle parts, information technology (IT) products, medical devices, and retreaded tires remanufacturing sectors. In addition, a case study on "other remanufacturing sectors" presents information for the consumer products, electrical apparatus, locomotives, machinery, office furniture, and restaurant equipment remanufacturing sectors.

[3] This report focuses on U.S. commercial remanufacturing activities. The Commission did not collect data on remanufacturing activities performed by the U.S. armed services.

[4] With the exception of trade in retreaded tires, the United States does not maintain statistics on trade in remanufactured goods.

[5] A copy of this questionnaire is presented in appendix E.

[6] For details on the Commission's survey methodology, see appendix F.

[7] The beginning of the 2009–11 study period coincides with the tail end of the 2008–09 U.S. economic recession and ensuing period of weak economic growth. The discussion in this report notes cases in which the effect of the recession on data trends merits attention.

[8] For the purposes of the Commission's questionnaire, firms that stated that they disassembled cores into parts or cleaned, inspected, and tested parts, but did not produce remanufactured goods, were classified as performing remanufacturing activities other than production. In addition, questionnaire respondents that stated that they sold, traded, warehoused, or distributed either cores or remanufactured goods, but did not produce remanufactured goods, were classified as also performing wholesaling remanufacturing activities. Only questionnaire respondents that stated that they produced remanufactured goods were classified as producers.

[9] The questionnaire defined a "foreign remanufacturer invested in the United States" as a firm that reported (1) that one or more foreign entities had a substantial (10 percent or greater) ownership interest in it as of December 31, 2011; and (2) that at least one of these foreign entities had remanufacturing activities in countries other than the United States. An "entity" is a corporation, limited-liability company, partnership, or sole proprietorship, or any subsidiary of any of these forms of businesses.

[10] For a list of hearing participants, see appendix C.

[11] For summaries of positions of interested parties, see appendix D.

[12] Australia, Brazil, Canada, Chile, China, all 27 European Union (EU) members, India, Indonesia, Japan, Malaysia, Mexico, New Zealand, Russia, Singapore, Korea, Taiwan, and Vietnam.

[13] Australia, Austria, Belgium, Brazil, Bulgaria, Canada, China, Cyprus, Czech Republic, Denmark, Estonia, Finland, France, Germany, Hungary, India, Korea, Lithuania, Malaysia, Malta, Mexico, the Netherlands, New Zealand, Poland, Portugal, Romania, Russia, Singapore, Spain, Sweden, Taiwan, the United Kingdom, and Vietnam.

[14] However, in certain sectors remanufacturing services may be performed where the core is located. For instance, in the electrical apparatus sector, generators are often remanufactured on site. In the aerospace sector, aircraft are often flown directly to the facility that provides maintenance services.

[15] One exception is retreaded tires, which are classified in NAICS (2007) code 326212.

[16] Hauser and Lund, The Remanufacturing Industry: Anatomy of a Giant, 2003, 6. The estimates were featured in Hauser and Lund, The Remanufacturing Industry: Hidden Giant, 1996.

[17] These included (in alphabetical order) the automotive, compressors, electrical apparatus, machinery, office furniture, tires, toner cartridges, and valves sectors, as well as an "all other" sector. Hauser and Lund, *The Remanufacturing Industry: Anatomy of a Giant*, 2003, 25. Product areas were defined by the Standard Industry Classification System (SIC), the precursor to the NAICS.

[18] Hauser and Lund, *Remanufacturing: Operating Practices and Strategies*, 2008, 7. In particular, original estimates of the number of firms, production, and employment in remanufacturing were based on trade associations' estimates of the number of nonmember firms within a given industry sector. In hindsight, the authors reported that they believed that the estimates of nonmember firms provided by these trade associations were vastly overstated. Ibid., 7.

[19] Giuntini and Gaudette, "Remanufacturing: The Next Great Opportunity for Boosting U.S. Productivity," 2003, 41.

[20] These included the agriculture; communications; computer; construction; electric generation; entertainment; heating, ventilation, and air conditioning (HVAC); instruments; manufacturing; medical; mineral extraction; office; printing and paper; transaction (e.g., payment processing equipment); and transportation (aircraft, maritime, locomotive, trucks, and tires) sectors, as well as an "other" sector. Ron Giuntini (principal, Giuntini and Co.), telephone interview with USITC staff, August 4, 2011.

[21] Ron Giuntini (principal, Giuntini and Co.), telephone interview with USITC staff, August 4, 2011. According to OPI, the mineral extraction equipment sector was the largest remanufacturing sector, accounting for 19 percent ($17.2 billion) of remanufacturing expenditures, followed by the aircraft sector (13 percent or $11.6 billion), the electric generation sector (10 percent or $9.2 billion), the automotive/light trucks sector (9 percent or $8.6 billion), and the manufacturing sector (8 percent or $7.9 billion).

[22] Based on USITC staff calculations of weighted responses to the Commission's questionnaire, machinery is the fourth-largest remanufacturing sector in terms of production value. However, the sector is broad, cuts across a variety of industries, and covers a diverse range of products, limiting its effectiveness as a distinct case study. As a result, quantitative and qualitative information on the machinery sector is presented together with other remanufacturing sectors described in chapter 9.

[23] Selected APEC and G-20 economies highlighted in chapter 10 of this report include Brazil (G-20), China (APEC and G-20), the EU (G-20), India (G-20), Korea (APEC and G-20), and Singapore (APEC). To a lesser extent, the chapter also provides information on Canada (APEC and G-20), Malaysia (APEC), Mexico (APEC and G-20), New Zealand (APEC), Russia (APEC and G-20), and Vietnam (APEC).

End Notes for Overview of U.S. Remanufacturing Activities

[1] McKenna, written testimony to the USITC, February 28, 2012, 1; Gager, written testimony to the USITC, February 28, 2012, 1.

[2] For a description of the Commission's survey methodology, see appendix F. For more information on remanufacturing data by size of firm, see tables G.1–G.9 in appendix G.

[3] Estimates of U.S. sales of remanufactured goods are based on USITC staff calculations of weighted responses to the Commission questionnaire. Estimates of total sales of all manufactured products in these industry sectors are based on U.S. Census data.

[4] In order to avoid double-counting, the Commission's questionnaire asked remanufacturers to report the percentage of their 2011 production shipments of remanufactured goods that include remanufactured goods made by other companies. The share of production value attributable to remanufactured intermediate inputs varies by sector. For instance, remanufactured intermediate inputs are estimated to have contributed 22 percent ($328 million) to the production of remanufactured medical devices in 2011, but only 11 percent ($305 million) to the production of remanufactured IT products that year. Overall, remanufactured intermediate inputs made by other companies are estimated to have contributed almost 16 percent ($6.7 billion) to U.S. production of remanufactured goods in 2011. As a result, estimates of U.S. production of remanufactured goods in 2011 may be somewhat overstated. For tables showing the percentage of U.S. remanufacturers that use remanufactured intermediate inputs, and the value share of those inputs by sector, see tables G.10 and G.11, respectively, in appendix G.

[5] For more information on the remanufactured HDOR equipment, machinery, and retreaded tires sectors, see chapters 4, 9, and 8, respectively.

[6] As noted in chapter 1 and appendix F, Commission estimates of the value of U.S. imports of remanufactured goods are based on the statistical analysis of imports reported in the Commission's questionnaire by firms that identified themselves either as producers of remanufactured goods or as performing remanufacturing activities, such as importing remanufactured goods. Other non-remanufacturing firms may directly import remanufactured goods. As a result, estimates of U.S. imports may be somewhat understated.

[7] Questionnaire respondents were asked to report the share of 2011 U.S. shipments of remanufactured goods sold to the U.S. armed services, including the Army, Navy, Marines, and Air Force (or reserve elements of the foregoing), as well as the Coast Guard and the National Guard. The Commission did not collect data on remanufacturing activities performed by the U.S. armed services.

[8] Based on USITC staff calculations of weighted responses to the Commission questionnaire.

[9] Questionnaire respondents were asked in question 8.1 to indicate the effect of various factors on their ability to compete in the U.S. market, and whether a given factor was "extremely important" to their competitive position.

[10] Based on USITC staff calculations to weighted responses to question 8.1 of the Commission questionnaire.

[11] Based on USITC staff calculations of weighted responses to the Commission questionnaire and on official statistics of the USDOC. For a listing of NAICS (2007) codes that correspond to these industry sectors, see table F.1, appendix F.

[12] Wholesaler trade includes remanufactured goods from other remanufacturing sectors. For instance, U.S. exports of remanufactured HDOR equipment by wholesalers are classified as wholesaler exports and not as U.S. exports of HDOR equipment. As a result, sector exports are likely somewhat understated. However, on a combined basis, sector and wholesaler exports reflect total U.S. exports of remanufactured goods.

[13] Questionnaire respondents were asked if one or more foreign entities had a substantial (10 percent or greater) ownership interest in their firm as of December 31, 2011 (an entity includes a corporation, limited liability company, partnership, sole proprietorship, or any subsidiary related to these forms of business). If a questionnaire respondent answered affirmatively, and stated that at least one of these entities had remanufacturing activities in countries other than the United States, then it was considered to be a foreign remanufacturer invested in the United States.

[14] USITC, hearing transcript, February 28, 2012, 88 (testimony of Brian A. Lewallen, Detroit Diesel Remanufacturing (DDR)).

[15] For leading factors affecting U.S. remanufacturers by sector, see table G.13 in appendix G.

[16] USITC, hearing transcript, February 28, 2012, 38–39, 88, and 96 (testimony of Greg Foley, Caterpillar Inc.; Brian Lewallen, DDR; and Allen Pierce, Cummins Inc.).

[17] The following U.S. FTAs (or trade promotion agreements) contain provisions addressing trade in remanufactured goods: Australia, Bahrain, CAFTA-DR, Chile, Colombia, Korea, Morocco, Oman, Panama, Peru, and Singapore. The Commission did not collect separate trade data for Colombia, Korea, and Panama as individual FTA partners because these agreements were not entered into force during the 2009–11 study period.

[18] The Commission collected and analyzed data on direct investment flows in U.S. and foreign remanufacturing activities. Data do not include direct investment stocks in remanufacturing activities invested in a U.S. or foreign firm prior to 2009.

End Notes for Aerospace

[1] For the purposes of this report, aerospace firms that produce remanufactured goods or provide remanufacturing services are considered to be remanufacturers. Likewise, firms that provide remanufacturing services are considered to ultimately produce remanufactured goods.

[2] Other aircraft include balloons, blimps, gliders, and rockets.

[3] The airframe is the main structure of an aircraft that contains all other systems. The airframe includes the wings, tail section, cabin, and fuel tanks. Butler, "Product Group Report," May, 2009, 7.

[4] Butler, "Product Group Report," May 2009, 7.

[5] Wheels and brakes are a major category of component maintenance, because of the significant wear and tear they endure during landings and the unique lightweight, heat-dissipating materials used. AeroStrategy, "Global MRO Market Economic Assessment," August 21, 2009, 14–15.

[6] Title 14 of the Code of Federal Regulations governs aviation safety. Under part 1.1 of 14 C.F.R., aircraft maintenance is defined as "inspection, overhaul, repair, preservation, and the replacement of parts, but excludes preventative maintenance." Thus, maintenance includes both overhaul and repair activities.

[7] See appendix A for the request letter from the USTR.

[8] Butler, "Product Group Report," May, 2009, 6. A rotable is a component or inventory item that can be repeatedly and economically restored to a fully serviceable condition. BusinessDictionary.com, http://www.businessdictionary.com/definition/rotable.html (accessed May 16, 2012).

[9] AeroStrategy, "Global MRO Market Economic Assessment," August 21, 2009, 2.

[10] Industry representative, telephone interview by USITC staff, July 19, 2012.

[11] AeroStrategy, "Global MRO Market Economic Assessment," August 21, 2009, 10, 18, 27, and 34.

[12] Based on USITC staff calculations of weighted responses to the Commission questionnaire. Certain aircraft parts and assemblies can be remanufactured to updated specifications that exceed the original specifications of the parts or assemblies. Whether or not to remanufacture the part to the original or updated specifications is at the discretion of the aircraft owner.

[13] Based on USITC staff calculations of weighted responses to the Commission questionnaire.

[14] Most commercial aircraft use gas turbine engines. By far, the aircraft engine represents the greatest expense of any component on an airplane, in terms of both initial outlay and maintenance costs. Butler, "Product Group Report," May 2009, 7.

[15] On the other end of the spectrum, Southwest Airlines has chosen to outsource 100 percent of its engine overhaul activities. Ang, "What's Hot," April 5, 2012.

[16] AeroStrategy, "Global MRO Market Economic Assessment," August 21, 2009, 14–15.

[17] AeroStrategy, "Global MRO Market Economic Assessment," August 21, 2009, 6–7.

[18] Ibid., 22–23.

[19] Based on USITC staff calculations of weighted responses to the Commission questionnaire.

[20] Michaels, "Seven Paradigm Shifts Reshaping Air Transport MRO," April 3, 2012.

[21] In addition to having one of the oldest airline fleets, the United States is also a world leader in domestic passenger traffic, accounting for just under a trillion revenue passenger kilometers (RPKs) in 2010. This total was about double that of the next area, Europe (intra-European travel was just under 500 million RPKs). Airfleets, *Airline Fleet Age* database (accessed June 18, 2012); USDOT, FAA, "Forecast of 2010 Top 10 Traffic Flows," March 9, 2012.

[22] Michaels, "Seven Paradigm Shifts Reshaping Air Transport MRO," April 3, 2012; Doan, "The Global MRO Forecast 2012–2022," April 3, 2012.

[23] For example, one industry analyst predicts that over the next 10 years, the Asia-Pacific and Middle East MRO markets will grow by 6.2 and 7.7 percent, respectively, while the North American and European markets will grow by 0.6 and 1.9 percent, respectively. Michaels, "Seven Paradigm Shifts Reshaping Air Transport MRO," April 3, 2012; Doan, "The Global MRO Forecast 2012–2022," April 3, 2012.

[24] Boeing predicts the North American regional and large civil aircraft fleet will grow from 6,610 to 9,330 by 2030, eclipsing Europe's projected fleet of 8,010 and China's projected fleet of 5,930. Boeing, *Current Market Outlook, 2011–30*, 2011. Airbus, however, predicts the Asia/Pacific fleet will be preeminent, at 10,398, followed by the North American fleet and European fleet. Airbus, *Global Market Forecast, 2011– 30*, 2011, 81, 86, and 93.

[25] Boeing forecasts a doubling of the world's airline fleet, from 19,410 in 2010 to 39,530 in 2030. Boeing, *Current Market Outlook, 2011–30*, 2011. Note: Boeing includes regional jets in its forecast, while Airbus does not. Airbus projects the market for aircraft with more than 100 seats to grow from 15,000 to just over 31,000 airliners during this period. Airbus, *Global Market Forecast 2011–30*, 2011, 60.

[26] In particular, composite repair and avionics repair will present challenges. Industry representative, telephone interview by USITC staff, Washington, DC, July 19, 2012.

[27] Boeing, "Boeing Edge Sets New Standard for Aviation Services and Support," February 15, 2012.

[28] Boeing, "787 GoldCare: Airplane Business Solutions" (accessed May 1, 2012). Industry sources indicate that as much as 60 percent of out-of-warranty work is managed by the OEMs, who typically subcontract parts of the job to other repair shops. Industry representative, telephone interview with USITC staff, July 19, 2012.

[29] Swearingin, "Customers Speak Out," April 4, 2012; Ang, "What's Hot," April 5, 2012.

[30] Based on USITC staff calculations of weighted responses to the Commission questionnaire and on official statistics of the U.S. Department of Commerce (USDOC) for NAICS (2007) codes 336411, 336412, and 336413.

[31] Based on USITC staff calculations of weighted responses to the Commission questionnaire and on official statistics of the USDOC for NAICS (2007) codes 336411, 336412, and 336413.

[32] Industry representative, telephone interview by USITC staff, Washington, DC, April 19, 2012.

End Notes for Heavy-Duty and Off-Road (HDOR) Equipment

[1] Based on USITC staff calculations of weighted responses to the Commission questionnaire.

[2] Industry representative, telephone interview with USITC staff, May 23, 2012.

[3] USITC, hearing transcript, February 28, 2012, 22 (testimony of Matthew P. Hehl, Remanufacturing Industries Council (RIC)).

[4] Industry representative, email message to USITC staff, January 27, 2012.

[5] Based on USITC staff calculations of weighted responses to the Commission questionnaire.

[6] USITC, hearing transcript, February 28, 2012, 167–68 (testimony of Donald R. Flatau, John Deere Reman, and Brian A. Lewallen, DDR).

[7] USITC, hearing transcript, February 28, 2012, 136–38 (testimony of Brian A. Lewallen, DDR; Allen W. Pierce, Cummins Inc.; and Donald R. Flatau, John Deere Reman).

[8] USITC staff calculations of weighted responses to the Commission questionnaire.

[9] MacKay and Company estimated the size of the broader U.S. market for remanufactured goods used in vehicular industries, which include light-, medium-, and heavy-duty trucks; agriculture equipment; and construction equipment. The U.S. market for remanufactured goods used in vehicular industries was estimated to be $15.2 billion in 2011, or about 25 percent of the $60 billion U.S. vehicular aftermarket. MacKay and Company, written submission to the USITC, February 2012. For more information on the remanufactured motor vehicle parts sector, see chapter 5.

[10] USITC, hearing transcript, February 28, 2012, 136–138 (testimony of Brian A. Lewallen, DDR).

[11] Industry representatives, interviews by USITC staff, Washington, DC, October 26, 2011; telephone interviews by USITC staff, July 18, 2012.

[12] Industry representatives, telephone interviews by USITC staff, July 19, 2012.

[13] Industry representatives, telephone interviews by USITC staff, July 17–19, 2012.

[14] Caterpillar Inc., "Cat Remanufactured 3054 Engine," 2004; Bosch Rexroth Corp., "Reman Exchange Program," n.d.; industry representative, telephone interview by USITC staff, July 19, 2012.

[15] Diesel and Gas Turbine Worldwide, "Remanufacturing vs. Rebuilding," March 2012, 10.

[16] Low-precision estimate; RSE below 65 percent.

[17] USITC, hearing transcript, February 28, 2012, 153–56 (testimony of Allen W. Pierce, Cummins Inc.; Brian A. Lewallen, DDR; and Donald R. Flatau, John Deere Reman).

[18] Based on USITC staff calculations of weighted responses to the Commission questionnaire and on official statistics of the USDOC for NAICS (2007) code 3331 (excluding 333112) and 333995.

[19] Based on USITC staff calculations of weighted responses to the Commission questionnaire and on official statistics of the USDOC for NAICS (2007) code 3331 (excluding 333112) and 333993.

[20] U.S. Customs ruling HQ 561465 (March 21, 2000) states that Caterpillar exports cores to Mexico for remanufacturing, and imports remanufactured goods into the United States. Similarly, U.S. Customs ruling HQ 547088 (August 29, 2002) states that Cummins does the same. See also the Caterpillar, Cummins, and Remy International websites.

[21] Figures for U.S. imports of remanufactured HDOR equipment by foreign remanufacturers invested in the United States were suppressed to prevent the disclosure of CBI.

[22] Figures for U.S. exports of HDOR cores by foreign remanufacturers invested in the United States were suppressed to prevent the disclosure of CBI.

[23] Figures for U.S. imports of HDOR cores by foreign remanufacturers invested in the United States were suppressed to prevent the disclosure of CBI.

[24] USITC, hearing transcript, February 28, 2012, 39, 95–96, and 149 (testimony of Greg Folley, Caterpillar Inc.; Allen W. Pierce, Cummins Inc; and Donald R. Flatau, John Deere Reman).

[25] USITC, hearing transcript, February 28, 2012, 87, 102–3, and 165–66 (testimony of Brian A. Lewallen, DDR; Donald R. Flatau, John Deere Reman; and Allen W. Pierce, Cummins Inc.).

[26] USITC, hearing transcript, February 28, 2012, 95–96 (testimony of Allen W. Pierce, Cummins Inc.).

[27] USITC, hearing transcript, February 28, 2012, 179 (testimony of Allen W. Pierce, Cummins Inc.).

[28] USITC, hearing transcript, February 28, 2012, 174–75 (testimony of David Fulghum, MacKay & Company).

[29] Based on USITC staff calculations of weighted responses to the Commission questionnaire.

[30] Muzinic, "SRC Integrates Newly Acquired Lexington Plant," May 3, 2010.

[31] USITC, hearing transcript, February 28, 2012, 82, 98, and 158 (testimony of Greg Folley, Caterpillar Inc.; Donald R. Flatau, John Deere Reman; and Allen W. Pierce, Cummins Inc.).

[32] USITC, hearing transcript, February 28, 2012, 82 (testimony of Greg Folley, Caterpillar Inc.).
[33] Caterpillar, "Caterpillar Expands Sustainable Product Offerings in Asia-Pacific," May 24, 2011.

End Notes for Motor Vehicle Parts

[1] This category includes parts for motor vehicles (i.e., passenger cars, light trucks, and medium- and heavy-duty on-road trucks). For information on retreaded tires, see chapter 8.
[2] The aftermarket comprises parts that are replacements for original equipment installed on new motor vehicles, as well as accessories and specialty equipment, such as floor mats and decals.
[3] McKenna, written testimony to the USITC, February 28, 2012, 2.
[4] IBISWorld Industry Report 42114, *Used Car Parts Wholesaling in the US*, March 2011, 19.
[5] USITC, hearing transcript, February 28, 2012, 32–33 (testimony of Rob Wright, Undercar Express).
[6] Ibid., 31 (testimony of Rob Wright, Undercar Express).
[7] APRA, "What is Remanufacturing?" (accessed February 16, 2012).
[8] Based on USITC staff calculations of weighted responses to the Commission questionnaire.
[9] Gager, written testimony to the USITC, February 28, 2012, 1.
[10] McKenna, written testimony to the USITC, February 28, 2012, 1; Gager, written testimony to the USITC, February 28, 2012, 1.
[11] For more information on the HDOR equipment remanufacturing sector, see chapter 4.
[12] IBISWorld Industry Report 42114, *Used Car Parts Wholesaling in the US*, March 2011, 13.
[13] The terms "vehicle makers" and "automakers" are used interchangeably throughout this chapter.
[14] Lund and Hauser, *Remanufacturing: An American Perspective*, Boston University.
[15] Industry representatives, interviews with USITC staff, April 26–27 and May 9, 2012.
[16] Industry representative, interview with USITC staff, May 9, 2012.
[17] Based on USITC staff calculations of weighted responses to the Commission questionnaire.
[18] Industry representative, telephone interview by USITC staff, July 30, 2012.
[19] Meritor, "Meritor's Environmentally Friendly Remanufacturing Efforts Promote Fleet Efficiency," June 28, 2012.
[20] USITC, hearing transcript, February 28, 2012, 47 (testimony of Robert McKenna, MEMA).
[21] Based on USITC staff calculations of weighted responses to the Commission questionnaire.
[22] USITC, hearing transcript, February 28, 2012, 44 (testimony of Rob Wright, Undercar Express).
[23] Ibid., 64–65 (testimony of Rob Wright, Undercar Express); Sevart, "Certified Transmission Thrives with New Distributors and Adds 100 Employees," March 2012, 6.
[24] Industry representative, interview with USITC staff, April 27, 2012.
[25] Shephardson, "Average Age of Cars, Trucks Hits 10.8 years, Sets Record," January 18, 2012.
[26] Shephardson, "Average Age of Cars, Trucks Hits 10.8 years, Sets Record," January 18, 2012; Goebel, "The Changing Auto Industry," March–April 2009, 7.
[27] Industry representative, telephone interview with USITC staff, March 29, 2012.
[28] Industry representative, interview with USITC staff, May 9, 2012.
[29] APRA, "What is Remanufacturing?"
[30] Industry representative, interview with USITC staff, April 27, 2012.
[31] USITC, hearing transcript, February 28, 2012, 64 (testimony of Robert McKenna, MEMA).
[32] Industry representative, interview with USITC staff, May 9, 2012.
[33] Clottey and Benton, "Core Acquisitions Planning," 2010, 3–4.
[34] Industry representative, interview with USITC staff, April 26, 2012.
[35] Industry representative, interview with USITC staff, April 27, 2012.
[36] Ibid.
[37] Industry representative, interview with USITC staff, April 26, 2012.
[38] Core brokers are middlemen in the core supply chain, and serve as a consolidation center for cores. They purchase cores from peddlers that purchase cores from vehicles at salvage operators, OEM/OES surplus, and dismantlers and then consolidate, sort, and link the cores to customer part numbers. Prices are negotiated, and the cores are then purchased and shipped to the remanufacturer. Schinzing, "Cores—Cores—Cores," Electronic Remanufacturing Company, July 2007.
[39] Industry representative, interview with USITC staff, May 9, 2012. In general, a core deposit is an additional charge incorporated into the price of a remanufactured good. Once a core is returned to the remanufacturer, the

deposit is credited back to the customer that returned the core. Such a system encourages core returns to ensure an adequate and reliable source of core supply.

[40] Industry representative, telephone interview by USITC staff, March 29, 2012; USITC, hearing transcript, February 28, 2012, 136 (testimony of Allen W. Pierce, Cummins Inc.).

[41] Industry representatives, interviews with USITC staff, April 26 and May 9, 2012.

[42] Sevart, "Certified Transmission Thrives with New Distributors," March 2012, 6.

[43] Ibid.

[44] Ibid.

[45] Based on USITC staff calculations of weighted responses to the Commission questionnaire and on official statistics of the USDOC for NAICS (2007) codes 333618 and 3363.

[46] Based on USITC staff calculations of weighted responses to the Commission questionnaire and on official statistics of the USDOC for NAICS (2007) codes 333618 and 3363.

[47] Low-precision estimate; RSE below 100 percent.

[48] Ibid.

[49] USITC, hearing transcript, February 28, 2012, 88 (testimony of Brian A. Lewallen, DDR).

[50] For an overview of the global markets for remanufactured goods, including an overview of measures affecting remanufacturers, see chapter 10.

[51] McKenna, written testimony to the USITC, February 28, 2012.

[52] USITC, hearing transcript, February 28, 2012, 178 (testimony of Allen W. Pierce, Cummins Inc.).

[53] USDOC, ITA, "On the Road: U.S. Automotive Parts Industry," 2010.

[54] Industry representative, interview with USITC staff, October 17, 2011.

[55] Meritor, "Meritor's Environmentally Friendly Remanufacturing Efforts," June 28, 2012.

[56] USITC, hearing transcript, February 28, 2012, 55 (testimony of Robert McKenna, MEMA).

[57] USITC, hearing transcript, February 28, 2012, 40 (testimony of Greg Folley, Caterpillar Inc.).

[58] USITC, hearing transcript, February 28, 2012, 47 (testimony of Robert McKenna, MEMA.

[59] AWTEC website, http://www.awtec.com/about (accessed June 18, 2012).

[60] Schau, Traverso, Lehmann, and Finkbeiner, "Life Cycle Costing in Sustainability Assessment," November 18, 2011, 2281.

[61] DDR website, http://www.detroitdieselreman.com/about/facilities.asp (accessed March 9, 2012); Cummins, "Remanufacturing is the New 'New'," 2011; Kekre et al., "Reconfiguring a Remanufacturing Line at Visteon, Mexico," November–December 2003, 30–43.

[62] Cardone website, http://www.cardone.com/about-us/locations-directions (accessed March 9, 2012); Remy International website, http://www.remyinc.com/business_structure.asp (accessed March 9, 2012).

[63] USITC, hearing transcript, February 28, 2012, 166 (testimony of Allen W. Pierce, Cummins Inc.).

[64] Cummins, "Cummins Megasite Opens in India," January 2011.

End Notes for Information Technology (IT) Products

[1] For the purposes of this report, cellular phones, including smartphones, are considered consumer products and not included in this chapter. For more information on consumer products, see chapter 9.

[2] Based on USITC staff calculations of weighted responses to the Commission questionnaire.

[3] Judge, written testimony to the USITC, February 9, 2012, 2.

[4] Industry representative, telephone interview by USITC staff, November 2, 2011.

[5] Based on USITC staff calculations of weighted responses to the Commission questionnaire.

[6] Ibid.

[7] Industry representative, telephone interview by USITC staff, April 27, 2012.

[8] Industry representatives, telephone interviews by USITC staff, November 1–2, 2011, and April 25, 2012.

[9] Based on USITC staff calculations of weighted responses to the Commission questionnaire.

[10] Judge, written testimony to the USITC, February 9, 2012, 3–4.

[11] Industry representative, telephone interview by USITC staff, November 1, 2011.

[12] Industry representative, telephone interview by USITC staff, November 2, 2011.

[13] Bernstein Research, "Remanufactured Cartridges," July 2005.

[14] Based on USITC staff calculations of weighted responses to the Commission questionnaire.

[15] Marcotte, Hallé, and Montreuil, "Computer Hardware Reverse Logistics," August 2008, 13.

[16] Industry representatives, telephone interviews by USITC staff, November 2, 2011, and February 2, 2012.

[17] Low-precision estimate; RSE below 65 percent.

[18] Industry representative, telephone interview by USITC staff, November 28, 2011.

[19] Industry representative, telephone interview by USITC staff, November 28, 2011. Under the act, warranty provisions that "require a purchaser of the warranted product to buy an item or service from a particular company to use with the warranted product in order to be eligible to receive a remedy under the warranty" are prohibited unless the company can demonstrate that the product will not work properly without that item or service. U.S. Federal Trade Commission (FTC), Bureau of Consumer Protection website, http://business.ftc.gov/documents/bus01-businesspersons-guide-federal-warranty-law#Magnuson-Moss (accessed April 30, 2012).

[20] Based on USITC staff calculations of weighted responses to the Commission questionnaire and on official statistics of the USDOC for NAICS (2007) codes 333313, 3341, 334290, 3344, 325910, 325992, 333293, and 333315.

[21] Based on USITC staff calculations of weighted responses to the Commission questionnaire and on official statistics of the USDOC for NAICS (2007) codes 333313, 3341, 334290, 3344, 325910, 325992, 333293, and 333315.

[22] Industry representatives, telephone interviews by USITC staff, November 2, 2011, and February 2, 2012.

[23] Marcotte, Hallé, and Montreuil, "Computer Hardware Reverse Logistics," August 2008, 13.

[24] Figures for U.S. imports of remanufactured IT products by foreign remanufacturers invested in the United States were suppressed to prevent the disclosure of CBI.

[25] Low-precision estimate; RSE below 80 percent.

[26] For more information on the global market for remanufactured goods, including China, see chapter 10.

[27] Government of Vietnam, Ministry of Information and Communications, *Circular Promulgating A List*, December 30, 2009.

[28] Industry representative, telephone interview by USITC staff, November 1, 2011. RoHS is an EU directive which states that "electrical and electronic equipment containing more than agreed levels of lead, cadmium, mercury, hexavalent chromium, polybrominated biphenyl and polybrominated diphenyl ether flame retardants" may not be placed on the EU market. UK Department for Business Innovation and Skills, "What is RoHS?" (accessed April 30, 2012).

[29] Low-precision estimate; RSE below 65 percent.

[30] Judge, written testimony to the USITC, February 9, 2012, 3–4.

[31] Respondents to the Commission questionnaire did not provide enough usable data to estimate inbound or outbound FDI in IT product remanufacturing.

End Notes for Medical Devices

[1] COCIR, Good Refurbishment Practice for Medical Imaging Equipment, October 2009, 5.

[2] The FDA defines a remanufacturer as "any person who processes, conditions, renovates, repackages, restores, or does any other act to a finished device that significantly changes the finished device's performance or safety specifications, or intended use." See 21 C.F.R. § 830.3.

[3] The FDA defines a reprocessor of SUDs as anyone who performs remanufacturing operations on an SUD. FDA, "Who Must Register, List and Pay a Fee," n.d. The reprocessing of SUDs reportedly accounts for the vast majority of FDA-defined remanufacturing. Industry representative, telephone interview by USITC staff, November 10, 2011.

[4] For the definition of remanufactured goods provided by the USTR, see chapter 1. Namely, refurbished, remanufactured, and reprocessed medical devices are generally disassembled, processed, cleaned, inspected, tested, and restored to original working condition or better, and are issued a warranty.

[5] Barlow, "Shaking Off a Stigma," April/May 2006.

[6] FDA-defined remanufactured medical devices other than reprocessed SUDs are thought to account for a very small portion of the U.S. market. As a result, remanufactured medical devices refer largely to refurbished medical imaging equipment and SUDs throughout the remainder of this chapter.

[7] USITC, hearing transcript, February 28, 2012, 192 (testimony of Michael Schmit, GE Healthcare Remanufacturing).

[8] IAMERS, Responsible Handling of Pre-Owned Medical Diagnostic Imaging Equipment, n.d.

[9] AMDR website (accessed May 16, 2012). For a complete list of commonly reprocessed devices, see http://www.amdr.org/industry-info/commonly-reprocessed-medical-devices/.

[10] COCIR, Good Refurbishment Practice for Medical Imaging Equipment, October 2009, 10.

[11] USITC, hearing transcript, February 28, 2012, 192 (testimony of Michael Schmit, GE Healthcare Remanufacturing).

[12] GE Healthcare, for instance, began refurbishing its devices through its GoldSeal program within the past eight years, while Siemens established its Proven Excellence refurbishing program in 2000. More recently, Toshiba unveiled its Assurance Refurbishment Program in 2009. GE Healthcare, "GoldSeal Refurbished Systems," n.d. (accessed August 30, 2012); Siemens, "Refurbished Systems," n.d. (accessed November 15, 2011); ITN, "Toshiba Introduces Toshiba Assurance Program," November 10, 2009; Barlow, "Shaking Off a Stigma," April/May 2006.

[13] WHO, Medical Device Regulations, 2003, 28.

[14] Barlow, "Shaking Off a Stigma," April/May 2006.

[15] Ibid.

[16] AMDR, prehearing written submission to the USITC, February 28, 2012, 4.

[17] Pyrek, "Reprocessing of Medical Devices," February 1, 2003; industry representative, interview with USITC staff, Washington, DC, May 4, 2012; industry representative, telephone interview with USITC staff, May 21, 2012.

[18] Industry representative, interview with USITC staff, Washington, DC, May 4, 2012; industry representative, telephone interview by USITC staff, May 9, 2012.

[19] Industry representative, interview with USITC staff, Washington, DC, November 10, 2011; Pyrek, "Reprocessing of Medical Devices," February 1, 2003.

[20] Remanufactured medical imaging equipment likely accounts for 80–85 percent of production of remanufactured medical devices, as U.S. production of remanufactured SUDs was reported to be $250– $300 million in 2011. AMDR, prehearing written submission to the USITC, February 28, 2012, 4.

[21] Based on USITC staff calculations of weighted responses to the Commission questionnaire.

[22] An estimated 46 percent of medical device remanufacturers experienced no change in production capacity, while an estimated 3 percent experienced a decline in production capacity. USITC staff calculations of weighted responses to the Commission questionnaire.

[23] MPO, "Lawmaker Proposes Repealing Controversial Device Tax," May 2010.

[24] The prevailing system of reimbursement was based on a "fee for service" model where various medical services were reimbursed regardless of the outcome of the procedure. A value-driven model proposes to associate reimbursements with improved patient health. Barnes, "Moving Away from Fee-forService," May 7, 2012.

[25] USITC, hearing transcript, February 28, 2012, 194 (testimony of Michael Schmit, GE Healthcare Remanufacturing) and 198 (testimony of Hans Beinke, Siemens Medical Solutions).

[26] The U.S. government is the country's largest insurer, reimbursing nearly two-thirds of domestic hospital admissions. S&P, Healthcare: Products & Supplies, February 4, 2010.

[27] Medical Devices Today, "Reimbursement Execs," January 5, 2008.

[28] PRWeb, "Global Medical Imaging Equipment," April 5, 2012.

[29] Martin et al., Growth in US Health Spending, 2012, 213.

[30] USITC, hearing transcript, February 28, 2012, 200 (testimony of Hans Beinke, Siemens Medical Solutions).

[31] Beinke, written testimony to USITC, February 9, 2012.

[32] PR Newswire, "The Market Outlook for Refurbished Medical Devices to 2016," January 16, 2012; Beinke, written testimony to the USITC, February 9, 2012; USITC, hearing transcript, February 28, 2012, 199 (testimony of Hans Beinke, Siemens Medical Solutions).

[33] IbisWorld, Medical Device Manufacturing in the U.S., June 2011; ITN, "Importance of Refurbished Equipment," November 1, 2010.

[34] Barlow, "Shaking Off a Stigma," April/May 2006.

[35] USITC, hearing transcript, February 28, 2012, 200 (testimony of Hans Beinke, Siemens Medical Solutions); Barlow, "Shaking Off a Stigma," April/May 2006.

[36] For instance, the technology for X-rays (an imaging device) has not significantly changed over the past decade. Barlow, "How Second-hand Equipment Can Mean First-rate Buys," June 11, 2007.

[37] Elario, "Remanufacturing Policy Roundtable," January 24, 2011.

[38] The typical life cycle of a medical imaging device is 10 years. Equipment is generally leased for the first 5–6 years and may then be returned to the OEM. The OEM may then either sell the device "as-is" to a third-party

distributor, refurbish the device itself, or dispose of the equipment. Industry representative, interview with USITC staff, Washington, DC, December 21, 2011.

[39] Twenty of the 30 largest global medical device manufacturing firms are based in the United States. MPO, "The Top 30," July/August 2011.

[40] Industry representative, telephone interview by USITC staff, May 9, 2012.

[41] Based on USITC staff calculations of weighted responses to the Commission questionnaire. Due to the cost of imaging equipment, devices are most commonly leased. The OEM retains ownership of the device for the duration of the contract unless the device is paid for by the end user, who in that case can also eventually trade in the equipment for newer technology. Industry representative, interview with USITC staff, Washington, DC, May 9, 2012.

[42] USITC, hearing transcript, February 28, 2012, 206 (testimony of Dan Vukelich, AMDR).

[43] For instance, in 2011, 86 hospital mergers or acquisitions occurred, which was the highest number over the past decade. Pyrek, "Reprocessing of Medical Devices," February 1, 2003; Levin Associates, "Health Care M&A Spending Rises," January 25, 2012; Dworkin, "Health Costs: Behind the Numbers," May 6, 2012.

[44] Public Law 111-148.

[45] Berry, "Mergers between Insurers and Hospitals Expected to Accelerate," October 25, 2010.

[46] Based on USITC staff calculations of weighted responses to the Commission questionnaire and on official statistics of the USDOC for NAICS (2007) codes 334510, 334517, 339112, 339113, 339114, and 339115.

[47] Based on USITC staff calculations of weighted responses to the Commission questionnaire and on official statistics of the USDOC for NAICS (2007) codes 334510, 334517, 339112, 339113, 339114, and 339115.

[48] Beinke, written testimony to USITC, February 9, 2012; Gibson, "Business Trends in Shipping," February 2, 2012.

[49] Beinke, written testimony to USITC, February 9, 2012, 4.

[50] This is because OEMs typically source their own cores from the markets in which they are located. Beinke, written testimony to USITC, February 9, 2012.

[51] Figures for U.S. imports of medical device cores by foreign remanufacturers invested in the United States were suppressed to prevent the disclosure of CBI.

[52] USDOC, Global Import Regulations, May 2008. For more information on the global markets for remanufactured goods, see chapter 10.

[53] Of the 106 countries for which market access information is available, 85 are believed to allow the importation of remanufactured medical devices on "same as new" terms. USDOC, Global Import Regulations, May 2008.

[54] USDOC, Global Import Regulations, May 2008.

[55] There are additional proposals to prohibit imports of remanufactured medical devices that were not originally produced in Brazil. USITC, hearing testimony, February 28, 2012, 196 (testimony of Michael Schmit, GE Healthcare Remanufacturing). For more information on Brazil, see chapter 10.

[56] USDOC, Global Import Regulations, May 2008; Beinke, written testimony to USITC, February 9, 2012.

[57] USITC, hearing testimony, February 28, 2012, 208 (testimony of Dan Vukelich, AMDR).

[58] Ibid.

[59] Beinke, written testimony to the USITC, February 9, 2012.

[60] Philips, "Refurbished Equipment: The Diamond Select Program," n.d. (accessed August 31, 2012).

[61] Industry representative, interview with USITC staff, Washington, DC, November 15, 2011.

[62] ITN, "Toshiba Introduces Toshiba Assurance Refurbishment Program," November 10, 2009.

[63] Philips remanufactures ultrasound systems in Solingen, Germany, and various imaging technologies in Veldhoven, the Netherlands. Philips, "Refurbished Equipment: The Diamond Select Program," n.d. (accessed August 31, 2012).

[64] Singh, "GE Sets Up Facility in Bangalore," May 18, 2009.

[65] Beinke, written submission to the USITC, February 9, 2012.

[66] Dineen, "Healthymagination Investor Update," November 9, 2010.

[67] GE Healthcare, "GE Healthcare Opens First Plant in Brazil," July 22, 2010.

End Notes for Retreaded Tires

[1] Given the common industry acceptance of the terms "retreaded tires" and "casings," for purposes of this report, remanufactured tires will be referred to as retreaded tires or tire retreads, while cores will be referred to as casings.

[2] Off-the-road (OTR) tires are typically larger, heavier tires used on construction and agricultural equipment.

[3] USITC, hearing transcript, February 28, 2012, 116 and 147 (testimony of Marvin Bozarth, TIA). In the 1950s, there were 12,000 retread plants in the United States, the majority of which performed retreading operations on tires for passenger vehicles. Over time, the cost of equipment and number of molds required to retread multiple sizes of passenger vehicle tires became too prohibitive for all but a few sizes that may date back 10–15 years, resulting in only a handful of passenger vehicle tire retreaders. Nowadays, passenger vehicle tires are typically not designed with retreading in mind.

[4] Tire Retread & Repair Information Bureau website, http://www.retread.org (accessed August 31, 2012); USDOC, Census, *Tire Retreading*, 2007.

[5] *Modern Tire Dealer*, "The Top 100 Retreaders in the U.S.," April 2011, 24.

[6] Industry representative, email message to USITC staff, January 2012.

[7] USITC staff calculations of weighted responses to the Commission questionnaire.

[8] *Modern Tire Dealer*, "The Top 100 Retreaders in the U.S.," April 2011, 24.

[9] Based on USITC staff calculations of weighted responses to the Commission questionnaire. New passenger car tires account for the majority of U.S. tire sales. However, since passenger car tires are not typically retreaded, retreaded tires account for a significantly higher portion of total U.S. sales of non-passenger tires, which include truck, aircraft, and OTR tires.

[10] USITC, hearing transcript, February 28, 2012, 116 (testimony of Marvin Bozarth, TIA); *Modern Tire Dealer*, "The Top 100 Retreaders in the U.S.," April 2011, 21–29.

[11] The U.S. Census Bureau (Census) also collects statistics on the U.S. retreaded tire industry. According to Census, U.S. shipments of retreaded tires were $1.1 billion in 2010. In 2007, 5,361 U.S. production workers were employed in 523 retreading establishments. Seventy percent of the retread operations employed fewer than 100 workers, although some larger retread facilities employed 100– 250 workers. USDOC, Census, *Tire Retreading (NAICS 326212)*, 2007.

[12] USITC, hearing transcript, February 28, 2012, 113–14 (testimony of Marvin Bozarth, TIA).

[13] Based on USITC staff calculations of weighted responses to the Commission questionnaire.

[14] *Modern Tire Dealer*, "The Top 100 Retreaders in the U.S.," April 2011, 21–29.

[15] Passenger car tires are typically not retreaded. Passenger car tires account for the vast majority of tires recycled for purposes other than retreading. Approximately 75 percent of spent passenger car tires are used for fuel and grounded rubber applications, while about 15 percent go to land disposal. Rubber Manufacturers Association, "U.S. Scrap Tire Management Summary 2005–2009," October 2011.

[16] For instance, Federal Executive Order 13149 (65 Fed. Reg. 24607 (April 21, 2000)) mandated the use of retreaded tires on certain government vehicles. Regulations established by the U.S. Environmental Protection Agency (40 C.F.R. Part 253) require that all federal, state, and local government agencies and contractors that use federal funds purchase retreaded tires or tire retreading services to the maximum extent practicable.

[17] *Modern Tire Dealer,* "Everyone Scrambles for Usable Truck Tires," January 2012, 42.

[18] USDOC, Census Bureau, *Tire Retreading*, 2007.

[19] Ulrich, "A Case for Casings," April 2011, 21; USITC hearing transcript, February 28, 2012, 112–114 (testimony of Marvin Bozarth, TIA); *Modern Tire Dealer*, "Everyone Scrambles for Usable Truck Tires," January 2012, 42.

[20] Bridgestone Bandag Tire Solutions, "Bridgestone Bandag Tire Solutions Announces Campaign," March 21, 2011.

[21] USITC, hearing transcript, February 28, 2012, 121 (testimony of David Stevens, TRIB).

[22] *Modern Tire Dealer*, "The Top 100 Retreaders in the U.S.," April 2011, 24.

[23] *Modern Tire Dealer*, "Everyone Scrambles for Usable Truck Tires," January 2012, 42. Bridgestone (Bandag) accounted for about 42 percent of the truck tire retread market; Goodyear, 28 percent; Michelin, 23 percent; Marangoni, 4 percent; and others, 3 percent.

[24] *Modern Tire Dealer*, "2011 Was a Great Year for Retreading," January 17, 2012.

[25] Industry representative, telephone interview by USITC staff, August 15, 2012.

[26] USITC, hearing transcript, February 28, 2012, 114 (testimony of Marvin Bozarth, TIA).

[27] USITC, hearing transcript, February 28, 2012, 112–13 (testimony of Marvin Bozarth, TIA); TIA and TRIB, "Understanding Retreading," 2009.

[28] TRIB website, http://www.retread.org (accessed 2011-2012).

[29] USITC hearing transcript, February 28, 2012, 115 (testimony of Marvin Bozarth, TIA).

[30] Industry representative, email message to USITC staff, November 29, 2011. [31] *Modern Tire Dealer*, "The Top 100 Retreaders in the U.S.," April 2011, 24.

[32] USITC, hearing transcript, February 28, 2012, 121 (testimony of David Stevens, TRIB).

[33] USITC, hearing transcript, February 28, 2012, 115 (testimony of Marvin Bozarth, TIA).

[34] *Modern Tire Dealer*, "The Top 100 Retreaders in the U.S.," April 2011, 24; "Prepping for Growth," April 2012, 34.

[35] Marangoni Tread North America, "Marangoni Expands North American Operations," August 31, 2011.

[36] *Modern Tire Dealer*, "The Top 100 Retreaders in the U.S.," April 2011, 21–30.

[37] Continental, "Shipments Begin from Continental's Flat Tread Production Facility in Mexico," May 26, 2011; "First ContiLifeCycle Dealership Opens in U.S," September 23, 2011.

[38] *Modern Tire Dealer*, "The Top 100 Retreaders in the U.S.," April 2011, 21–30.

[39] In 2011, retreaded truck and bus tires accounted for about 43 percent of total U.S. retreaded tire exports; OTR construction and agricultural retreads, 39 percent; and passenger vehicle retreads, 18 percent.

[40] Official statistics of the USDOC for NAICS (2007) codes 326211 and 326212.

[41] Official statistics of the USDOC for NAICS (2007) codes 326211 and 326212.

[42] Respondents to the Commission questionnaire did not provide enough usable data to estimate inbound or outbound FDI in tire retreading activities.

[43] Goodyear, *10-K Annual Report*, 2011, 19.

End Notes for Other Remanufacturing Sectors

[1] For statistical purposes, information collected in the Commission's questionnaire from the electrical apparatus, locomotives, office furniture, and restaurant equipment sectors was aggregated to provide greater statistical significance; the tables therefore show figures for machinery, consumer products, and "all other."

[2] Industry representative, telephone interview by USITC staff, February 21, 2012.

[3] Based on USITC staff calculations of weighted responses to the Commission questionnaire and on official statistics of the USDOC. For a listing of NAICS (2007) codes that correspond to these industry sectors, see table F.1, appendix F.

[4] Based on USITC staff calculations of weighted responses to the Commission questionnaire and on official statistics of the USDOC. For a listing of NAICS (2007) codes that correspond to these industry sectors, see table F.1 in appendix F.

[5] Solar Turbines, Inc., "Turbotec Remanufacturing Facility," March 2009.

End Notes for Global Trade, Industries, and Markets for Remanufactured Goods

[1] As noted in chapter 1 of this report, information presented in this chapter responds to the eighth and final element of the USTR's request.

[2] APEC members are Australia (also G-20), Brunei Darussalam, Canada (G-20), Chile, China (G-20), Hong Kong, Indonesia (G-20), Japan (G-20), Korea (G-20), Malaysia, Mexico (G-20), New Zealand, Papua New Guinea, Peru, the Philippines, Russia (G-20), Singapore, Taiwan, Thailand, the United States (G-20), and Vietnam. Additional G-20 members include Argentina, Brazil, France, Germany, India, Italy, Saudi Arabia, South Africa, Turkey, and the EU.

[3] As discussed in this chapter, the EU's Directive on End-of-Life Vehicles and its Directive on Waste Electrical and Electronic Equipment (WEEE) set targets for recycling and reuse in the automotive and electrical/electronics sectors, respectively. By setting targets for the reuse of parts, these measures may encourage remanufacturing. On the other hand, the EU's Restriction on the Use of Hazardous Substances (RoHS) Directive prohibits the reuse of components containing certain substances, which may raise remanufacturing costs for some EU remanufacturers.

[4] Parker, *Remanufacturing in the UK*, 2004, 3, 68–70. The UK has the largest aerospace industry in the EU, and the second largest in the world, after the United States.

[5] Parker, An Analysis of the Spectrum of Re-use, 2007, 26.

[6] Lufthansa Technik established an MRO facility for passenger aircraft in Sofia, Bulgaria, in 2008. The company has repaired/refurbished 50 aircraft at this facility and plans to service 35 additional aircraft in 2012. U.S. Department of State, U.S. Embassy, Sofia, "Remanufactured Medical Equipment, Automotive Parts, and Airplanes," January 12, 2012.

[7] Grose, "Born Again," 2007, 53–55.

[8] Chapman et al., *Remanufacturing in the UK*, 2010, 17.

[9] According to the International Federation of Engine Rebuilders and Remanufacturers (FIRM), there are 900 companies that rebuild or remanufacture motor vehicle and truck engines in Europe. Lammel, written submission to the EC's Internet consultation, n.d. (accessed September 2, 2012).

[10] Chapman et al., *Remanufacturing in the UK,* 2010, 34.

[11] ETIRA website, "Frequently Asked Questions" (accessed September 19, 2011).

[12] Multinational firms engaged in the remanufacture of printer toner cartridges in Spain include Cartridge World (Australia), Hewlett-Packard (United States), and Katun (United States). U.S. Department of State, U.S. Embassy, Madrid, "Madrid Response to USITC," January 25, 2012. Leading remanufacturers of ink and toner cartridges in the UK include Danwood, Greenstrike, and Xerox. U.S. Department of State, U.S. Embassy, London, "United Kingdom: Remanufactured Goods Market," January 30, 2012. At least three producers of ink cartridges in France use remanufactured cartridges: Armor, MBP, and Pelikan. U.S. Department of State, U.S. Embassy, Paris, "Embassy Paris Response to USITC Request," January 18, 2012.

[13] U.S. Department of State, U.S. Embassy, London, "United Kingdom: Remanufactured Goods Market," January 30, 2012; Gray and Charter, "Remanufacturing and Product Design," 2007, 37. However, the company reportedly ceased remanufacturing activities and focused on document processing services at its Gloucestershire location. *ThisIsGloucestershire*, "Xerox Is Now a Giant PO Box for UK Firms," March 11, 2010.

[14] U.S. Department of State, U.S. Embassy, The Hague, "The Netherlands: Remanufactured Goods Market Very Limited," January 12, 2012. While based in the Netherlands, Océ is owned by Canon of Japan.

[15] ETIRA website, "Key Facts," (accessed June 22, 2012).

[16] Chapman et al., *Remanufacturing in the UK*, 2010, 27.

[17] Chapman et al., *Remanufacturing in the UK*, 2010, 31–32; U.S. Department of State, U.S. Embassy, The Hague, "The Netherlands: Remanufactured Goods Market Very Limited," January 12, 2012. Estimates indicate that between 15 and 20 percent of the medical equipment imported into Bulgaria is remanufactured. U.S. Department of State, U.S. Embassy, Sofia, "Remanufactured Medical Equipment, Automotive Parts, and Airplanes," January 12, 2012.

[18] Directive 2000/53/EC of the European Parliament and of the Council, September 18, 2000.

[19] Chapman et al., *Remanufacturing in the UK*, 2010, 18. The directive instructs member states to ensure that economic operators set up systems for the collection, treatment, and recovery of end-of-life vehicles and that the last holder/owner of an end-of-life vehicle not be assessed any fee for delivering a vehicle to an authorized treatment center. It also instructs members to establish regulations to encourage motor vehicles producers to design their vehicles to (a) avoid the use of carcinogenic heavy metals and (b) facilitate the reuse and recycling of components. Further, the directive instructs members to require manufacturers of motor vehicles to give customers instructions for dismantling the vehicle at the end of its useful life and for properly handling potential hazardous waste resulting from the dismantling process.

[20] Directive 2012/19/EU of the European Parliament and of the Council, July 4, 2012.

[21] Parker, An Analysis of the Spectrum of Re-use, 2007, 83.

[22] *M2Presswire*, "TAM Airlines Registers an Operating Profit," March 10, 2011.

[23] Sobie, "Brazil MRO Sector Poised for Major Expansion," July 6, 2010; Morrison, "How Portugal's TAP M&E Became Brazil's Biggest MRO," May 22, 2012.

[24] Industry representatives, email messages to USITC staff, January 10 and 16, 2012.

[25] Industry representative, email message to USITC staff, January 16, 2012.

[26] Industry representative, email message to USITC staff, January 16, 2012.

[27] Industry representative, email message to USITC staff, January 10, 2012.

[28] Ibid.

[29] Industry representative, email message to USITC staff, January 16, 2012.

[30] Ibid.

[31] Datamonitor, "Caterpillar Inc., Company Profile," September 20, 2011, 7.
[32] Datamonitor, "Komatsu Ltd., Company Profile," June 2, 2011, 5.
[33] Oliveria, "The State of the Remanufacturing Industry in Brazil," October 11, 2011.
[34] Datamonitor, "Lexmark International, Inc., Company Profile," August 17, 2011, 5.
[35] *Recharger Magazine,* "Reciclamais South American Expo Promotes Brazilian Market," February 1, 2008.
[36] Oliveria, "The State of the Remanufacturing Industry in Brazil," October 11, 2011.
[37] Datamonitor, "Lexmark International, Inc., Company Profile," 5.
[38] *Recharger Magazine,* "Reciclamais South American Expo Promotes Brazilian Market," February 1, 2008.
[39] With a few exceptions, the government prohibits the importation of used vehicles and used consumer products. U.S. Department of State, U.S. Embassy, Brasilia, "Brazil: Response to ITC Inquiry," February 12, 2012.
[40] Import licensing systems are administrative procedures requiring the submission of an application or other documentation (other than that typically required for customs purposes). Under non-automatic import licenses, certain conditions must be met before a license to import is issued. USTR, "Import Licensing," n.d. (accessed September 2, 2012).
[41] Applications for imports of used goods are published by Brazil's Foreign Trade Secretariat (SECEX) to allow domestic industries to comment within 30 days of the application to prove that a similar product is manufactured in Brazil. U.S. Department of State, U.S. Embassy, Brasilia, "Brazil: Response to ITC Inquiry," February 12, 2012.
[42] The age of the goods must be supported by (1) proof of year of manufacture; (2) proof of year of restoration to original working condition of the good and supporting documentation of specifications; (3) a list of technical differences between the inspected unit and a new unit of the same kind; (4) a declaration of the average useful life of the product; and (5) proof of the market value of the good in question, the reproduction value of a similar but new product, and the replacement value of a technologically updated equivalent product. U.S. Department of State, U.S. Embassy, Brasilia, "Brazil: Response to ITC Inquiry," February 12, 2012.
[43] U.S. Department of State, U.S. Embassy, Brasilia, "Brazil: Response to ITC Inquiry," February 12, 2012.
[44] Cummins prehearing submission, February 9, 2012.
[45] Ibid.
[46] For more information on the U.S. remanufactured medical device sector, see chapter 7.
[47] USDOC, *Global Import Regulations*, May 2008, 12–13.
[48] U.S. Department of State, U.S. Embassy, Brasilia, "Brazil: Response to ITC Inquiry," February 12, 2012.
[49] U.S. Department of State, U.S. Embassy, Brasilia, "Brazil: Response to ITC Inquiry," February 12, 2012.
[50] Ibid.
[51] Ibid.
[52] Geetha, "Remanufacturing Growth," n.d. (accessed September 2, 2012).
[53] Ibid.
[54] Geetha, "Remanufacturing Growth," n.d. (accessed September 2, 2012).
[55] ICRRA website, http://www.icrra.com (accessed July 24, 2012); CRTAI website, http://www.crtai.org (accessed July 24, 2012).
[56] USITC, hearing transcript, February 28, 2012, 158 (testimony of Allen Pierce, Cummins Inc.).
[57] NBM Media, "Volvo Outlines Growth Plans for India," May 2011.
[58] USITC, hearing transcript, February 28, 2012, 162, 175–76 (testimony of Allen Pierce, Cummins Inc.).
[59] Ibid., 175 (testimony of Allen Pierce, Cummins Inc.).
[60] Government of India, Ministry of Commerce and Industry, Department of Commerce, *Hand Book of Procedures (Vol. I)*, 2009–2014, effective August 27, 2009, article 2.33.
[61] Government of India, Ministry of Commerce and Industry, Department of Commerce, *Foreign Trade FTP, 2009–2014*, August 23, 2010, article 2.17. The publication, which is revised periodically, is the principal document that outlines India's policies governing imports and exports.
[62] USITC, hearing transcript, February 28, 2012, 42–43 (testimony of Greg Folley, Caterpillar Inc.).
[63] China's General Armament Department approved the creation of the laboratory, which is housed within the Academy of Armored Forces Engineering of the People's Liberation Army, most likely in support of the remanufacturing of military equipment. Located in Beijing, the academy is China's principal training and engineering campus for China's armored and mechanized military forces. The laboratory also funds research projects at various companies and sponsors industrywide conferences to promote the industry and improve remanufacturing knowledge throughout China. Sina.com.cn, "Remanufacturing: Looks Beautiful," March 31, 2012; Holeung Ho Lee Foundation, "Xu Binshi," n.d. (accessed September 2, 2012).

[64] These include the Law of the People's Republic of China on Cleaner Production Promotion, effective January 1, 2003, and the Law of the People's Republic of China on Prevention of Environmental Pollution Caused by Solid Waste, effective April 1, 2005.

[65] The Law of the People's Republic of China on Circular Economy Promotion, effective January 1, 2009, stipulated that the term "reusing" included remanufacturing. Specific provisions of the law require that remanufactured products must meet certain standards and that remanufactured products must bear a label stating that they are remanufactured: unlabeled remanufactured products will be subject to penalties.

[66] Prior to these two pilot programs, a few companies appear to have conducted remanufacturing operations with official permission. In 2005, Ecostar (Nanjing Tianzhong Electromechanical Remanufacturing Co., Ltd) began the reconditioning or remanufacturing of old, imported photocopiers as a trial pilot program. Jinan Fuqiang Power Co., Ltd., began remanufacturing motor vehicle engines and engine components in 2006. Ecostar, "State General Administration of AQSIQ Visited ECOSTAR," October 24, 2005; China National Heavy Duty Truck Group Jinan Fuqiang Power Co., Ltd., "About Us," n.d. (accessed September 2, 2012).

[67] Other government ministries that formulate policies that affect the development of China's remanufacturing industry include the General Administration of Quality Supervision, Inspection, and Quarantine (AQSIQ); the Ministry of Finance (MOF); the General Administration of China Customs (GACC); the Ministry of Public Security (MPS); and the Ministry of Commerce (MOFCOM). U.S. Department of State, U.S. Embassy, Beijing, "China's Remanufacturing Market," February 1, 2012.

[68] NDRC, "Guidelines for Promoting the Development," May 31, 2010.

[69] NDRC, "Notice on Experimental Work on Deepening Remanufacturing," September 14, 2011.

[70] U.S. Department of State, U.S. Embassy, Beijing, "China's Remanufacturing Market," February 1, 2012.

[71] Many of these products can also be classified as HDOR equipment. Remanufacturers in this sector can therefore include both motor vehicle parts and HDOR remanufacturers.

[72] Xiang and Ming, "Implementing Extended Producer Responsibility," 2011, 680–86.

[73] State Council, "Decree No. 307 of the State Council of the People's Republic of China: Measures for Administration of Recycling of Scrapped Motor Vehicles," Gazette of the State Council of the People's Republic of China, August 20, 2001.

[74] Chen and Zhang, "End-of-Life Vehicle Recovery in China," March 2009, 50.

[75] Government of China, MIIT, "Ministry of Industry and Information Technology Issues Notices," December 21, 2009.

[76] Caterpillar Inc., "Caterpillar Announces Remanufacturing Joint Venture," December 14, 2009.

[77] Caterpillar Inc., "Caterpillar in China: Shanghai," n.d.

[78] Fuji Xerox, Co., Ltd., "Chinese Government Certifies Fuji Xerox Recycling Site," February 2, 2010.

[79] IBM, "IBM Opens the First Server Remanufacturing Center in China," February 29, 2012.

[80] U.S. Department of State, U.S. Embassy, Beijing, "China's Remanufacturing Market," February 1, 2012.

[81] Government of China, MIIT and NDRC, *Automotive Industry Development Policy*, September 1, 2009.

[82] USITC, hearing transcript, February 28, 2012, 82–83 (testimony of Greg Folley, Caterpillar Inc.).

[83] USITC, hearing transcript, February 28, 2012, 82–83 and 158 (testimony of Greg Folley, Caterpillar Inc.; and Allen W. Pierce, Cummins Inc.).

[84] Government of China, MOFCOM, "Measures for Administration of Import," April 7, 2008; Government of China, MOFCOM, "Catalog for 2012 of Goods Subject to Import Licenses," December 31, 2011; Government of China, MOFCOM, GACC, and AQSIQ, "Import License Management Catalog of Imports 2012," December 30, 2011.

[85] For example, products classified in HS chapters 84, 85, and 90.

[86] USITC, hearing transcript, February 28, 2012, 196 (testimony of Michael Schmit, GE Healthcare Remanufacturing).

[87] Ibid.

[88] USTR, "2011 U.S.-China Joint Commission," November 21, 2011.

[89] *Retreading Business*, "Zhu Jun Explains the Current Status of the CTRA," issue 2010/1; *Retreading Business*, "Interview: China Tyre Retreading, Repairing and Recycling Association," issue 2009/2. In comparison, there are about 680 tire retreaders in the United States, ranging in size from small firms producing around 20 retreaded tires per day (7,500 tires annually) to large firms producing more than 350,000 tires per year. For more information on the U.S. retreaded tire sector, see chapter 8 of this report.

[90] *Retreading Business*, "Chinese Retread Volume to Increase by 30%," issue 2009/3.

[91] Ibid.

[92] *Retreading Business*, "Interview: China Tyre Retreading, Repairing and Recycling Association," issue 2009/2.

[93] *Retreading Business*, "Zhu Jun Explains the Current Status of the CTRA," issue 2010/1.

[94] Bridgestone Corp., "Bridgestone Officially Opens Aircraft Tire Retreading Plant in China," July 25, 2007.

[95] *Retreading Business*, "Interview: China Tyre Retreading, Repairing and Recycling Association," issue 2009/2.

[96] U.S. Department of State, U.S. Embassy, Singapore, "An Overview of the Remanufacturing Industry in Singapore," March 9, 2012.

[97] Caterpillar Inc., "Caterpillar Announces Opening of New Singapore Remanufacturing Facility," May 24, 2011.

[98] PR Newswire, "Caterpillar Inc.: APEC Forges Ahead on Remanufacturing," March 29, 2012.

[99] U.S. Department of State, U.S. Embassy, Singapore, "An Overview of the Remanufacturing Industry in Singapore," March 9, 2012; PRWeb, "Singapore Launches Asia's Inaugural Remanufacturing R&D Center," June 22, 2012.

[100] Information in this section is based on U.S. Department of State, U.S. Embassy, Seoul, "ROK Remanufacturing Market," March 9, 2012.

End Notes for Appendix D

[1] Note that page numbers for exhibits are not precise because the exhibits volume is not paginated sequentially—the page numbers are those given by Adobe PDF software.

[2] USITC, hearing transcript, February 28, 2012, 202–09; AMDR, written submission to the USITC, February 28, 2012. For more information about AMDR, see http://www.amdr.org.

[3] APRA, written testimony to the USITC, February 9, 2012. For more information about APRA, see http://www.buyreman.com.

[4] USITC, hearing transcript, February 28, 2012, 36–41; Caterpillar, written submission to the USITC, February 28, 2012; Caterpillar, posthearing submission to the USITC, March 8, 2012. For more information about Caterpillar Inc., see http://www.cat.com.

[5] USITC, hearing transcript, February 28, 2012, 91–96; Cummins, written submission to the USITC, February 9, 2012. For more information about Cummins Inc., see http:/www.cummins.com.

[6] USITC, hearing testimony, February 28, 2012, 183–191. For more information about Davies, see http://www.daviesoffice.com.

[7] USITC, hearing transcript, February 28, 2012, 97–106; DDR, written testimony to the USITC, February 28, 2012. For more information about DDR, see http://www.detroitdiesel reman.com.

[8] USITC, hearing transcript and exhibits, February 28, 2012, 9–16.

[9] USITC, hearing transcript, February 28, 2012, 49.

[10] USITC, hearing transcript, February 28, 2012, 52–53.

[11] EASA, written submission to the USITC, February 7, 2012. For more information about EASA, see http://www.easa.com.

[12] Written submission to the USITC, May 19, 2012. For more information about NEMA, see http://www.nema.org.

[13] USITC, hearing transcript, February 28, 2012, 192–97; GE, written submission to the USITC, February 28, 2012. For more information about GE, see http://www.gehealthcare.com.

[14] USITC, hearing transcript, February 28, 2012, 122–130; IITC, written submission to the USITC, February 9, 2012. For more information about IITC, see http://www.i-itc.org.

[15] USITC, hearing transcript, February 28, 2012, 96–103; John Deere Reman, written submission to the USITC, February 28, 2012. For more information about John Deere Reman, see http://www.deere.com/en_us/reman/.

[16] USITC, hearing transcript, February 28, 2012, 103–111; McKay & Company, written submission to the USITC, February 28, 2012. For more information about MacKay & Company, see http://www.mackayco.com.

[17] Medtronic, written submission to the USITC, May 15, 2012. For more information about Medtronic, see http://www.medtronic.com.

[18] MEMA, written submission to the USITC, February 28, 2012; USITC, hearing transcript, February 28, 2012, 23–29; MEMA, posthearing submission to the USITC, March 8, 2012. For more information about MEMA, see http://www.mema.org.

[19] Mr. McKenna testified on behalf of MEMA in conjunction with MERA.

[20] MEMA, posthearing written submission to the USITC, March 8, 2012.

[21] Ibid.

[22] USITC, hearing transcript, February 28, 2012, 209–213. For more information about PCRR, see http://www.pcrr.com.

[23] USITC, hearing transcript, February 28, 2012, 16–23; RIC, written testimony to the USITC, February 28, 2012. For more information about RIC, see http://www.remancouncil.org.

[24] USITC, hearing testimony, February 28, 2012, 213–221; RLA, written submission to the USITC, February 28, 2012. For more information about RLA, see http://www.reverselogisticstrends.com.

[25] USITC, hearing transcript, February 28, 2012, 197–202; Siemens, written submission to the USITC, February 9, 2012. For more information about Siemens, see http://www.medical.siemens.com.

[26] USITC, hearing transcript, February 28, 2012, 111–117. For more information about TIA, see http://www.tireindustry.org.

[27] USITC, hearing transcript, February 28, 2012, 117–122; TRIB, written submission to the USITC, February 28, 2012. For more information about TRIB, see http://www.retread.org.

[28] USITC, hearing transcript, February 28, 2012, 29–36; Undercar Express, LLC, written submission to the USITC, February 28, 2012. For more information about Undercar Express, LLC, see http://www.ucx.com.

[29] Undercar Express, LLC, is a member of MERA. Mr. Wright's testimony was also presented on behalf of MERA.

End Notes for Appendix F

[1] Remanufacturing intensity can also be defined as the ratio of the value of production shipments of remanufactured goods to total sales of all products within a given industry sector. Based on this measure, Commission staff calculated the remanufacturing intensity of these 12 industry sectors based on the statistical analysis of questionnaire results, as described in the executive summary and in chapters 2–9 of this report.

[2] The industry sectors are (in alphabetical order): aerospace, consumer products, electrical apparatus, heavy-duty and off-road (HDOR) equipment, information technology (IT) products, locomotive systems, machinery, medical devices, motor vehicles, office furniture, restaurant equipment, and tires.

[3] Commission staff limited the sample population to firms with 20 or more employees in order to reduce the population size to be sampled. Restricting the population count enables improved statistical properties, as a larger share of the relevant population can be sampled. Further, in this case, restricting the population to firms with 20 or more employees eliminated only a relatively small fraction of remanufacturing employment and production. For instance, according to the U.S. Census Bureau, establishments with fewer than 20 employees accounted for 64.4 percent of all establishments classified in remanufacturing-intensive sectors in 2007, but they accounted for only 4.4 percent of employment and 2.7 percent of the value of production shipments.

[4] Not counting duplicate entries, which were removed from the population size. Treatment of duplicate entries is described below.

[5] SMEs were sampled randomly based on a Neyman allocation method to determine how many firms in each stratum to choose. Rather than sample a constant rate (e.g., 20 percent) of each stratum, the Neyman allocation takes into account the heterogeneity of the population within each stratum to determine how many firms should be sampled. Strata that are very heterogeneous, as determined by the variance in employment across firms in each stratum, are sampled at relatively high rates, while strata that are relatively homogeneous are sampled at lower rates. This method yields results with better statistical properties than a constant sampling rate.

[6] In certain cases, firms may have been correct that the questionnaire did not apply to them, and in other cases firms mistakenly certified that the questionnaire did not apply to them—for example, firms that do not consider themselves to be remanufacturers even if they appear to produce goods that meet the definition of remanufactured goods provided to the Commission by the USTR.

[7] Sampling weights can therefore be defined as the inverse probability of inclusion within a stratum.

[8] For more information on the application of RSEs, see USDOC, Census, "How the Data Are Collected" (accessed August 22, 2012).

[9] Using Stata, a statistical software program, this can be done via the "singleunit" option when setting up the design elements of the survey database.

In: Remanufactured Goods
Editor: Lori A. Simmons

ISBN: 978-1-62618-501-2

Chapter 2

COMMENTS TO FEDERAL TRADE COMMISSION, RE: REQUEST FOR PUBLIC COMMENTS, USED AUTO PARTS GUIDE REVIEW*

Motor and Equipment Manufacturers Association (MEMA)

INTRODUCTION

The Motor & Equipment Manufacturers Association (MEMA) represents over 700 companies that manufacture and remanufacture motor vehicle parts for use in the light- and heavy-duty original equipment and aftermarket industries. Motor vehicle parts manufacturers are the nation's largest manufacturing sector, directly employing over 685,000 people across the country. MEMA represents its members through four affiliate associations: Automotive Aftermarket Suppliers Association (AASA); Heavy Duty Manufacturers Association (HDMA); Motor & Equipment Remanufacturers Association (MERA); and Original Equipment Suppliers Association (OESA).

MEMA, on behalf of its affiliate association, MERA, is pleased to submit comments in response to the Federal Trade Commission's (FTC) request for public comments regarding its revision to the "Guides for the Rebuilt, Reconditioned and Other Used Automobile Parts Industry" (Guides). In short, the Guides are outdated and outmoded because they suggest that remanufactured automotive products and various used automotive products are largely equivalent, thereby ignoring the significant advancements this country's remanufacturing industry has made with regard to consumer quality and in the area of international trade since the Guides were first adopted in 1979. Therefore, MEMA and MERA urge the Commission to overhaul the Guides to reflect these advancements and to encourage consumers to choose remanufactured products, which are produced using energy efficient and environmentally sound manufacturing processes. Such action would be consistent with other ongoing federal efforts.

---

* These comments to FTC Project No. P127702 were made August 3, 2012.

The motor vehicle parts industry has led the way in remanufacturing. Examples of remanufactured components in the industry include: engines, transmissions, alternators, starters, turbochargers, steering and suspension components, and electronic control modules.

The remanufacturing industry provides a sustainable and economically viable response to increased global trade pressures, resource scarcity and rising energy costs—all for the benefit of the U.S. consumer. Remanufacturing preserves the value of the original manufacturing—including material and energy costs and investments in capital and labor inputs—which recycling alone cannot do. Remanufacturing saves 85 percent of the energy and material used to manufacture equivalent new parts, making it a sustainable and economically viable response to increased global trade pressures, resource scarcity and rising energy costs.[1] Remanufactured parts are on average 20 to 50 percent below the cost of most new parts, while delivering an equivalent level of quality backed by very competitive warranties.[2]

In 1997, the Environmental Protection Agency (EPA) recognized the qual ity of remanufactured parts when it published a study titled "Remanufactured Products: Good as New."3 Since the remanufacturi ng process incorporates improvements in product design, in many cases a remanufactured product may perform better than the original part.

Remanufactured motor vehicle parts also represent an exceptional value for consumers. However, public awareness about remanufacturing and the corresponding economic, environmental and product performance benefits is all too lacking in the consumer marketplace.

M EM A and M ERA strongly support efforts to educate consumers on the significant role remanufactured motor vehicle parts play in vehicle maintenance and repair. The FTC Guides can certainly assist with this important effort. To do so, the FTC must:

1. A l ign with other government agencies, including the Office of the U nited States Trade Representative (USTR) and U.S. Customs and Border Protection, states and municipalities, and recognize that remanufactured motor vehicle parts, while not "new," should also not be classified as "used" parts;
2. Rename the Guides to reflect this important distinction; and,
3. Recognize that "remanufacturing" is a standardized industrial process—hence, remanufactured parts are produced in a factory setting.

REMANUFACTURED PARTS: NOT NEW, NOT USED

Although a remanufactured part is not "new," classifying it as "used" is confusing to consumers and inaccurate for i nternational trade purposes.

The Guides as written cover a broad range of automobile parts, from (re)used to remanufactured. That range also includes: repaired, refurbished, rebuilt, factory rebuilt and others. Refer to the attached chart, "Comparison of Used, Remanufactured and New Automobile Parts." In most cases, the term "used" implies an "as-is" status to consumers, which is the case with the majority of vehicles labeled as "used" under the FTC Used Car Rule. In the automotive aftermarket, "used" parts are also typically sold "as-is" and can be sourced from end-of-l ife vehicles in salvage yards. In many cases, these parts will adequately meet the needs of consumers.

Similarly, most consumers understand the concept of repair. A n existing part such as a throttle body assembly can be repai red through proper cleaning and adjustment without the replacement of component parts. If the failure of any existing part cannot be addressed through cleaning or adjustment, the component part can be replaced with one that is reused, repaired, rebuilt, remanufactured or new, which can create confusion for the consumer.

Rebuilding can include dismantling, cleaning and reconstruction of a worn or non-functional product.

And in many cases, it is tested for functionality. Although a rebuilt part may seem similar to a remanufactured part, there is a distinction.

MERA and others define "remanufactured" parts as those produced using a standardized industrial process by which previously sold, worn or non-functional products are returned to same-as-new, or better, condition and performance. The process incorporates technical specifications, including engineering, quality, and testing standards to yield *fully warranted* products."

This definition has also been adopted by the Remanufacturing Industries Council (RIC), which was founded nearly 20 years ago at the Rochester Institute of Technology in Rochester, New York. It is also very similar to the definition being used by CL EPA, the European Association of Automotive Suppliers. In fact, M ERA, RIC and CL EPA are engaged in an ongoing effort to harmonize a definition of remanufacturing that can be used across remanufacturing industries and across borders.

Remanufacturing goes beyond rebuilding by using a standardized industrial process to yield consistency and repeatability in the overall manufacturing and testing process. This provides the consumer with a product that is equivalent to—and in many cases better than—new and with a full warranty.

Consequently, it creates confusion for consumers to classify remanufactured parts—with qual ity consistent or superior to new—i n the same category as parts purchased from a salvage yard.

These changes are not only important for consumers, but also the U.S. economy. Testi mony during a February 28, 2012 ITC hearing on remanufacturing pointed to the intentional misuse of the term "used" by many international trading partners to create non-tariff barriers for U.S. exports of remanufactured goods.[4]

M E M A and M ERA urge the FTC to no longer classify "remanufactured" parts as "used" and to reference language in the recently-passed Colombia, Panama and South Korea Free Trade Agreements (FTAs) regarding remanufactured goods.

The followi ng definitions are used in Article 4.23 of the U.S.-Colombia FTA:[5]

> "remanufactured goods means industrial goods assembled in the territory of a Party classified under Harmonized System Chapter 84, 85, 87, or 90 or heading 94.02, except goods classified under Harmonized System heading 84.18 or 85.16, that:
>
> a) are entirely or partially comprised of recovered goods; and
>
> b) have a similar life expectancy and enjoy a factory warranty similar to such new goods;
>
> recovered goods means materials in the form of individual parts that are the result of: (a) the disassembly of used goods into individual parts; and (b) cleaning, inspecting, testing, or other processes as necessary for improvement to sound working condition."

RENAMING OF THE GUIDES

Once the Part 20 Guide is clarified to reflect the distinction that "remanufactured" is not "new" or "used," we urge the FTC to rename the Guides. Neither the official name ("Guides for the Rebuilt, Reconditioned and Other Used Automobile Parts Industry") nor the shortened name referenced in the Request for Comments ("Used Auto Parts Guide") accurately categorize remanufactured parts.

M E M A and M ERA urge the FTC to rename the Guides (recommendation: "Guides for the Used and Remanufactured Automobile Parts Industries").

REMANUFACTURED PARTS: PRODUCED IN A FACTORY SETTING

Remanufacturing utilizes a standardized industrial process; hence, remanufactured parts are produced in a factory setting. As such, consumers may be confused by the terms "remanufactured" and "factory rebuilt," as referenced in 16 C.F.R. §20.3. Additionally, because "remanufactured" is a more commonly used term than "factory rebuilt" in the motor vehicle parts industry, there is an opportunity to bring clarity to this section for consumers.

M E M A and M ERA urge the FTC to engage in further discussions with M E M A and M ERA on the wording for this section.

M E R A is currently developing an effort to elevate public awareness of the economic, environmental and product performance benefits of remanufacturing and would welcome the opportunity to partner with the FTC in this endeavor. In fact, the important role and growing impact of remanufacturing in the U.S. and global motor vehicle markets led to the creation of M ERA—the affiliate association of M EM A focused on remanufacturi ng—in 2011.

Among MERA's top priorities is an education and awareness campaign, named M anufactured AgainTM, to heighten the understanding and acceptance of remanufacturing in the motor vehicle parts industry.

The campaign will include a certification program that will let consumers and commercial customers know that remanufactured parts from M ERA member companies are truly remanufactured. The program will define minimum process criteria for products to be eligible for Manufactured AgainTM certification and include industry certifications such as ISO 9001 and TS 16 949.

To achieve M anufactured AgainTM certification, each remanufacturer must demonstrate that they utilize a standardized industrial process in a factory setting. Other criteria will apply, but the important distinction is that Manufactured AgainTM remanufacturers are first and foremost manufacturing companies.

The manufacturing processes they utilize consistently produce high-quality products that are fully warranted. This effort will lead to a process certification seal that can be affixed to the part and/or box and used in advertising and other promotional materials by participating companies. M EM A and M ERA urge the FTC to follow the advancement of this program, which will continue to educate consumers on the important benefits of remanufactured motor vehicle parts.

CONCLUSION

Both federal and state officials have begun to research and promote the benefits of remanufactured parts. Most notably, USTR requested an ITC investigation of the size and scope of the U.S. remanufacturi ng industry. The study, titled "Remanufactured Goods: A n Overview of the U U.S. and Global Industries, Markets and Trade,"[6] will be instrumental in further defining the motor vehicle parts remanufacturing industry by providing important data. The ITC will issue its report by October 28, 2012. In December 2011, a bipartisan coalition of M embers of Congress submitted a request to the U.S. Government Accountability Office (GAO) to study the use of remanufactured parts in the federal vehicle fleet.[7] Earlier this year, the GAO indicated that it will begin the necessary research to conduct the study. M EM A and M ERA expect the GAO to complete the study in 2013. M EM A and M ERA urge the FTC to review these reports, as their findings may be helpful when revising the Guides.

MEMA and MERA are not aware of federal, state or local laws that conflict with the Guides, but, in recent years, there has been legislative activity at the state level seeking to promote remanufacturing in the motor vehicle parts industry. We applaud these efforts as they will help inform consumers and expand remanufacturing in the marketplace.

Between 2007 and 2010, resolutions were passed in Pennsylvania on the state and local level to promote the use of remanufactured parts in the corresponding government vehicle fleets. Most recently, Michigan in 2011 approved consideration of remanufactured goods in its procurement process. State governments are recognizing the benefits of remanufacturing and this helps to rightfully legitimize the industry for the marketplace. The Guides could benefit by examining state efforts to enhance the role of the motor vehicle parts remanufacturing industry.

MEMA and MERA continue to work with USTR and the Department of Commerce to promote remanufactured goods trade with our trading partners, and we commend the government for including remanufactured goods in the recently passed Colombia, Panama and South Korea Free Trade Agreements. Government efforts to promote remanufacturing at the Asia-Pacific Economic Cooperation forum and in the current Trans-Pacific Partnership trade negotiations are important steps to increasing global trade in remanufactured goods that will support the domestic U.S. industry.

In closing, MEMA and MERA appreciate the opportunity to comment on the Guides and that the information provided will help improve the Guides' ability to promote fair competition and consumer awareness. We urge the FTC not to finalize the Guides in the current format, which does not properly recognize the significant advancements made by the U.S. remanufacturing industry over the past 30 years. Instead, the Commission should overhaul the Guides to reflect this ongoing evolution of the remanufacturing industry, and particularly to capture the federal government's other actions that recognize that remanufactured motor vehicle parts, while not "new," should not be classified as "used" parts.

In addition, MEMA and MERA request a meeting with the FTC to further discuss the industry and the process for revising the Guides. For further information, please contact MEMA Director of Government Relations Dan Houton at dhouton@mema.org or 202-312-9250.

Respectfully submitted,

Robert E. McKenna
President & CEO, MEMA

John R. Chalifoux
President & COO, MERA

End Notes

1 Nabil Z. Nasr, Newton B. Green II, David R. Fister, and Monica M. Becker, "Capturing Energy and Resources through Remanufacturing," Conference Proceedings, EUROPEAN COUNCIL FOR AN ENERGY EFFICIENT ENVIRONMENT (2003) (http://www eceee.org/conference proceedings/ACEEE industry/2003/ Panel 4/p4 21/paper).

2 Nabil Z. Nasr, Rochester Institute of Technology, "Remanufactured Goods: An Overview of the U.S. and Global Industries, M arket, and Trade, pg. 7 (Feb. 2012) (http://www.usitc. gov/press room/documents/testimony/332 525 001.pdf).

3 U.S. Environmental Protection Agency, Publication No. EPA530-N-97-002 (May 1997) (http://www.epa. gov/osw/conserve/smm/wastewise/pubs/wwupda6.pdf).

4 "Remanufactured Goods: An Overview of the U.S. and Global Industries, Markets, and Trade," Hearing Before U.S. International Trade Comm'n, Feb. 28, 2012 (http://www. usitc.gov/press room/spotlight /remanufactured goods testi mony.htm).

5 Article 4.23 of the U.S.-Colombia Free Trade Agreement (http://www.ustr.gov/sites/default/ files/uploads/ agreements/fta/colombia/asset upload file788 10154.pdf).

6 77 Fed. Reg. 11157 (Feb. 24, 2012) (https://www.federal regi ster.gov/articles/2012/02/ 24/2012-4262/ remanufactured-goods-an-overvi ew-of-the-us-andglobal-i ndustries-markets-and-trade-change-i n-start).

7 Dec. 12, 2011, Letter from 18 M embers of Congress to Eugene Dodaro, Comptroller General, Government Accountabi l ity Office (available at http://www.mera.org/Document-Vault/GAO-Reman-Request-FINAL.pdf).

In: Remanufactured Goods
Editor: Lori A. Simmons

ISBN: 978-1-62618-501-2

Chapter 3

USE OF REMANUFACTURED PARTS IN THE FEDERAL VEHICLE FLEET IS BASED ON A VARIETY OF FACTORS*

United States Government Accountability Office

In fiscal year 2011, federal civilian agencies reported about $975 million in maintenance and repair costs for approximately 588,000 vehicles that the agencies owned.[1] These vehicles can be maintained and repaired using new or remanufactured parts. While there is no standard definition of a remanufactured vehicle part, the Federal Acquisition Regulation defines remanufactured parts as factory rebuilt to original specifications.[2] Remanufactured vehicle parts tend to be less expensive than comparable new parts.

The principal remanufactured products in the motor vehicle sector are engines, transmissions, starter motors, alternators, steering racks, and clutches, according to the U.S. International Trade Commission.[3]

Given the potential for cost savings from using remanufactured parts in the federal fleet, you asked us to examine this issue. Thus, this report describes (1) the vehicle repair process for selected agencies, including the use of remanufactured parts, and (2) the factors that agency officials consider when deciding whether to use new or remanufactured parts for repairs.

To perform our work, we reviewed the General Services Administration (GSA), the U.S. Postal Service (USPS), and the Departments of Agriculture (USDA), Homeland Security (DHS), the Interior (DOI), and Justice (DOJ). These six federal entities accounted for 95 percent of the civilian federal fleet vehicles in 2011. Because the four departments decentralize fleet management to their component agencies, we selected three agencies within each department with large fleets. In total, we selected 14 agencies for our review:

- GSA;
- USPS;

---

* This is an edited, reformatted and augmented version of The United States Government Accountability Office publication, GAO-13-316R, dated March 7, 2013.

- USDA
  - U.S. Forest Service,
  - Animal and Plant Health Inspection Service, and
  - Natural Resources Conservation Service;
- DHS
  - Customs and Border Protection,
  - Immigration and Customs Enforcement, and
  - Federal Emergency Management Agency (FEMA);
- DOI
  - Fish and Wildlife Service,
  - National Park Service, and
  - Bureau of Land Management; and
- DOJ
  - Federal Bureau of Investigation (FBI),
  - Drug Enforcement Administration, and
  - U.S. Marshals Service.

We reviewed laws, regulations and agency documents related to vehicle fleet management, maintenance, and repair, including the Federal Management Regulation on Motor Vehicle Management; fleet management manuals and guidance of individual federal agencies; and the Federal Fleet Report for fiscal year 2011,[4] which summarizes fleet data submitted by agencies and reported to Congress.

To gain a better understanding of the remanufacturing industry, we attended an October 2012 conference on vehicle parts remanufacturing sponsored by the Motor and Equipment Remanufacturers Association (MERA) and toured the factory of a MERA member that remanufacturers alternators, starters, and generators. To obtain information from agency and commercial garages, we conducted site visits at a USPS maintenance facility, a GSA Maintenance Control Center (MCC) in Texas, and a commercial facility that services GSA-leased vehicles. At one MCC, we interviewed officials and listened to calls between GSA repair technicians and vendors who were requesting purchase orders for government vehicles that had been brought for repair or maintenance. We chose these sites based on their proximity to GAO offices.

The information collected from these interviews and a non-probability sample of site visits is not generalizable to all agency maintenance facilities or commercial garages but provided insights about the process of making vehicle repair decisions. We interviewed officials responsible for the vehicle fleet at the 14 selected agencies. We conducted this performance audit from August 2012 to March 2013 in accordance with generally accepted government auditing standards.

Those standards require that we plan and perform the audit to obtain sufficient, appropriate evidence to provide a reasonable basis for our findings and conclusions based on our audit objectives.

We believe the evidence obtained provides a reasonable basis for our findings and conclusions based on our audit objectives.

SUMMARY

The 14 agencies we reviewed have vehicle repair processes that generally allow field office staff, such as the vehicle operator or local fleet manager, to make vehicle maintenance and repair decisions. These processes neither mandate nor prohibit the use of remanufactured parts. The agencies we reviewed generally do not keep data on the extent to which remanufactured parts are used. Agency officials from the 14 agencies in our review said that they use remanufactured parts when warranted, and we observed the presence of remanufactured parts in the stock rooms during our visits to repair facilities. Deciding when to use remanufactured parts, according to agency officials and related guidance, depends on a number of factors including the cost, availability, and reliability of the part. According to agency officials, these factors are considered on a case-by-case basis for each repair to yield the best value for the federal government.

Table 1. Characteristics of the Largest Civilian Agency-Owned Vehicle Fleets for Fiscal Year 2011

Selected agencies	Agency-owned fleet inventory[a]		Average age[b]
GSA Fleet-leasing program	199,729		3.5[c]
U.S. Postal Service	209,684		16.3
Department of Agriculture	36,183		5.2
U.S. Forest Service		*18,559*	
Natural Resources Conservation Service		*9,184*	
Animal and Plant Health Inspection Service		*4,742*	
Department of Homeland Security	48,917		4.0
Customs and Border Protection		*27,216*	
Immigration and Customs Enforcement		*13,274*	
Federal Emergency Management Agency		*1,062*	
Department of Justice	38,787		4.4
Federal Bureau of Investigation		*18,968*	
Drug Enforcement Administration		*6,974*	
U.S. Marshals Service		*4,573*	
Department of the Interior	23,844		6.8
Fish and Wildlife Service		*6,860*	
National Park Service		*6,839*	
Bureau of Land Management		*3,902*	
Total selected agencies	557,144		
Remaining civilian agencies	31,137		
Total	588,281		

Source: GAO analysis of 2011 Federal Fleet Report and data reported to FAST by agencies.

[a] Inventory for USPS and the four departments includes only agency-owned vehicles, not commercially leased and GSA-leased vehicles. The GSA Fleet inventory includes vehicles leased to both civilian and military agencies. We included the vehicles leased to military agencies because the vehicles are owned by GSA.

[b] Average age is only for agency-owned vehicles, not the agency's commercially leased or GSA-leased vehicles.

[c] This number is approximate. The actual average age is 3.45 but we rounded to 3.5 years.

BACKGROUND

Despite the common use of the term "federal fleet," there is no single entity responsible for managing all of the approximately 588,000 vehicles that federal civilian agencies own. Instead, each agency is responsible for managing its own fleet.

Agencies may own vehicles or lease them from GSA Fleet[5] or the commercial sector. Although some federal agencies, such as USPS, may have a single, centralized fleet of vehicles, the fleets of other agencies, such as the four departments in our review, are decentralized and dispersed among multiple sub-agencies, bureaus, or divisions. See table 1 for more information on each of these agencies' fleets.

GSA's Office of Government-wide Policy establishes fleet management policies for the federal government and issues guidance to help agencies manage their fleets effectively and meet federal requirements. GSA issues guidance primarily through its *Guide to Federal Fleet Management*. According to the *Guide*, "the challenge of any fleet maintenance process is to mix the principal requirements for achieving fleet maintenance—facilities and equipment, mechanic labor, parts, cross-sharing, and commercially contracted or outsourced services—in order to maximize vehicle reliability, safety, availability, and operating performance while minimizing labor, parts, and contracted service expenditures." Decisions about vehicle repair are one aspect of the fleet maintenance challenge.[6]

VEHICLE REPAIR DECISIONS ARE DECENTRALIZED AND MAY INCLUDE USE OF REMANUFACTURED PARTS

The fleet management policies and guidance of the 14 selected agencies generally give local field offices the decision-making responsibilities for vehicle maintenance and repair. Decision-making within the repair process is typically delegated to the vehicle operator, local field office, or fleet manager. Departmental and agency-level policies provide guidance to help local officials manage their fleets; however, the policies neither mandate nor prohibit the use of either remanufactured or new replacement parts.

Field staff of the selected agencies generally follow a common repair process for their agency-owned vehicles[7]—a process similar to the one individuals follow when repairing a personally-owned vehicle, as described below. A major decision point within the repair process is tied to whether a vehicle is covered by the manufacturer's warranty. Also, field staff may have the opportunity to choose whether new or remanufactured parts are used.

- Identify vehicle problems: When a vehicle operator has a problem with the assigned vehicle, the operator takes it to a garage, either an in-house, agency-owned garage or a commercial garage. A mechanic may also find a problem during regularly scheduled preventive maintenance or other repairs.
- Vehicle warranty in effect: If the vehicle is under the manufacturer's warranty, the field staff are to generally take the vehicle to a dealership or a garage that is authorized to perform warranty work in order to maintain the vehicle's warranty and have the work performed at no cost to the government. Manufacturer's warranties specify what types of repairs are covered. Some warranties allow the dealerships and

authorized garages to use remanufactured parts for repairs. For example, one agency provided us with the warranties for three of its vehicles, and we found all of them allowed the use of remanufactured parts.

- Vehicle warranty has expired: If the vehicle manufacturer's warranty has expired, the garage is to determine what repairs are needed and give the operator a repair estimate. If the garage is in-house, the mechanic or other agency staff may choose new or remanufactured replacement parts.8 If a commercial garage is used for repair, the garage may provide field staff the opportunity to choose a new or remanufactured replacement part.
- Approval of repair work: The operator or a supervisor reviews the estimate and approves the repair work. Agencies impose thresholds on the dollar amount of repair estimates, which, if exceeded, must be approved by a supervisor or another agency official. The threshold varies among agencies. For example, commercial garages servicing GSA Fleet vehicles call GSA's MCC for approval of repairs over $100. The Drug Enforcement Administration's supervisory approval threshold is $250. DHS's Immigration and Customs Enforcement and FEMA, for example, require supervisory approval for repairs over $2,500, and some other agencies do so for repairs over $3,000.
- The garage performs the repair, returns the vehicle to the operator, and the agency pays the bill.

Some agencies have in-house garages, such as USPS, FBI, and DHS's Customs and Border Protection; however, they may use commercial garages for some maintenance and repair work. Other agencies, such as GSA, use commercial garages for all, or almost all, of their maintenance and repair needs. For agencies that use both in-house and commercial garages for vehicle repairs, the decision to use a commercial or in-house garage is affected by factors such as the proximity of the vehicle to an in-house garage, the workload of the in-house garage, and whether a vehicle is under the manufacturer's warranty.

Agency officials from all 14 agencies told us that they use remanufactured parts to maintain and repair their vehicle fleets, including starters, alternators, brake calipers,[9] engines, and transmissions. While the agencies track maintenance costs for vehicles in their inventory, officials from 13 agencies—all except GSA—said they do not keep data on the extent to which they use remanufactured parts. GSA officials stated that GSA Fleet tracks the number of remanufactured parts used, but its information system understates the number because commercial vendors do not always indicate if a part is remanufactured. According to GSA officials, GSA is working to improve its data collection efforts by reviewing the information system's coding list to ensure its accuracy. Officials also stated that GSA is requiring additional training for its MCC employees to help ensure that when they enter data on repairs, they capture information on the use of remanufactured parts. Officials at the GSA MCC we visited told us that remanufactured parts, if used, were included in the coding list under "rebuilt."[10]

The opportunity to use more remanufactured parts increases as vehicles age and vehicle parts begin to wear out. This opportunity differs significantly for the two largest fleets in our review—USPS and GSA Fleet—which collectively account for about 70 percent of the vehicles owned by civilian agencies. As was shown in table 1, USPS, with close to 210,000 vehicles in 2011, has the oldest average age of vehicles at 16.3 years.[11] According to officials, USPS uses many remanufactured parts to maintain and repair its vehicles, particularly transmissions and engines. Figure 1 shows an example of a remanufactured transmission being removed from a common USPS vehicle; it is to be replaced with another remanufactured transmission. On the other hand, GSA officials stated that because GSA Fleet's 200,000 vehicles are relatively young, averaging about 3.5 years, they require few repairs and rarely require expensive repairs, such as replacing engines or transmissions.

Agencies Consider Multiple Factors When Deciding Whether to Use New or Remanufactured Parts

Agency officials from the 14 selected agencies stated that field staff consider a number of factors when deciding whether to use new or remanufactured parts and do so on a case-by-case basis for each repair to yield the best value for the federal government. [12]

Source: GAO.

Figure 1. A Removed Remanufactured Transmission at a USPS Vehicle Maintenance Facility.

The factors include:

- *Availability of the part:* Some parts may be immediately available while others could take days to arrive, a situation that could delay the vehicle's return to service and result in down-time for agency field staff.

- *Costs:* Agency officials said they consider the cost of the replacement part, the cost to ship it, and the cost of labor to install it. If the cost difference between a new and remanufactured part is small and the remanufactured part needs to be shipped, it could ultimately make the remanufactured part more expensive than a new part that does not require shipping.
- *Expected remaining useful life of the vehicle:* Agency officials said that they try to spend the least amount possible for a vehicle's repair if they know that they will soon dispose of the vehicle.
- *Reliability of the part:* Agencies rely on mechanics to use their personal experiences and knowledge to help inform their decisions on choosing the brands of remanufactured parts with which they have had the greatest success in the past.
- *Length of warranty:* The length of warranty for both new and remanufactured parts vary and may influence which replacement part is chosen.

Agency officials and commercial vendors provided several examples of using these factors to determine which parts to purchase. Cost of the part, of course, drives many decisions. In one example, a commercial vendor that works on GSA Fleet and other federal agency vehicles stated that when replacing an alternator for a Bluebird passenger bus, a remanufactured alternator cost about $850 and came with a 12-month/12,000-mile warranty, whereas a new alternator cost about $1,300 with a similar warranty. The commercial vendor chose the remanufactured alternator to repair the vehicle because of the part's lower cost and comparable warranty. In another example, a National Park Service official offered this scenario: If the Service needed an alternator for one of its fleet trucks and had a choice of either a remanufactured part for about $135 with a non-transferable lifetime warranty or a new part that the vehicle manufacturer made for about $426 with the manufacturer's warranty, the Service would choose the remanufactured part because it offered an acceptable warranty and significant cost savings. In yet another example, we listened to a telephone conversation in which a GSA MCC technician instructed a commercial vendor who was repairing a 2007 truck to use a remanufactured starter that cost about $206, rather than a new starter that cost about $360.

While several agency officials told us that the cost of a replacement part is an important factor in determining which type to use, they also stated that there are instances in which other factors, such as the cost of labor, part availability, remaining vehicle useful life, and part reliability can be more important. For example, officials from three agencies—USPS, GSA, and FBI—explained that if the installation of a relatively inexpensive part is expected to be labor-intensive, new parts are sometimes selected to avoid the cost associated with the risk of reinstalling the part if the remanufactured part fails prematurely. FEMA officials provided an example in which they stated that part availability is an important factor if the vehicle is critical to FEMA's mission or is a one-of-akind asset. If it takes three weeks to get a remanufactured part, but a new part is readily available, FEMA officials said they would purchase the new part, even if it were more expensive, to get the vehicle operational and back on the road as soon as possible. In an example related to remaining useful vehicle life, FBI officials discussed a repair scenario in which one of its vehicles had an expected remaining useful life of 6 years and needed its water pump replaced. In this case, they said that they would consider the replacement part's warranty and cost. If a new water pump cost $92 and had a lifetime warranty, while a remanufactured water pump cost $30 and had a 3-year warranty, then FBI officials would most likely select the new, more expensive water pump because of its lifetime warranty, extensive labor time associated with the repair, and the FBI's plan to use that vehicle for 6 more years. In an example of part reliability's importance as a factor, a GSA MCC technician and a commercial vendor agreed to replace an alternator in a 2003 passenger bus with a remanufactured alternator because remanufactured alternators for these vehicles are considered more reliable than new alternators. While remanufactured parts tend to be less expensive than comparable new parts, some agency officials told us they have found instances in which new parts are less expensive than remanufactured parts or are very close in price. FBI officials provided us an example of a new alternator with a lifetime warranty that cost $154 and a remanufactured alternator with a 3-year warranty that cost $179. In another instance, Fish and Wildlife Service officials told us that they decided to use a new alternator when they found that it cost only $5 more than the remanufactured alternator; the $5 savings from using the remanufactured part would be lost to shipping fees once the replaced alternator was mailed back to the remanufacturer.[13]

AGENCY COMMENTS

We provided a draft of this report to GSA, USPS, USDA, DOI, DHS, and DOJ for their review and comment. USPS and DOI provided written comments expressing their agreement with the findings of our report. USPS reiterated that it relies heavily on the remanufacturing industry to maintain its vehicle fleet. DOI stated that it actively seeks to control the costs of fleet maintenance and will continue to use remanufactured parts where they are the most cost-effective solution. GSA, USDA, DHS, and DOJ had no comments.

Lorelei St. James
Director
Physical Infrastructure Issues

End Notes

[1] Agencies annual maintenance and repair costs are published in the Federal Fleet Report. The General Services Administration provided GAO with its fiscal year 2011 maintenance and repair costs for the GSA Fleet.

[2] Federal Acquisition Regulation §52.211-5.

[3] U.S. International Trade Commission, Remanufactured Goods: An Overview of the U.S. and Global Industries, Markets, and Trade, Investigation No. 332-525, USITC Publication 4356 (October 2012).

[4] The Federal Fleet Report is updated annually and available on GSA's website at http://gsa.gov/portal/category/102859.

[5] GSA Fleet is a vehicle-leasing program for federal agencies, and maintenance and repair are managed by GSA.

[6] We are currently examining agencies' overall efforts to manage their vehicle fleets and expect to issue the results of our review later this year.

[7] For vehicles that agencies lease from GSA Fleet, GSA Fleet provides maintenance and repairs, among other services.

[8] According to agencies with in-house garages, with the exception of USPS, these garages do not stock inventories or they stock limited inventories that include items such as oil filters, brake pads, hoses, belts, and inventories may include remanufactured parts. However, these agencies generally obtain parts from available sources on an as needed basis. USPS's in-house garages maintain inventories that include both new and remanufactured parts.

[9] Brake calipers squeeze the brake pads against the surface of the brake rotor to slow or stop the vehicle.

[10] MCC officials stated that relatively few calls involve rebuilt or remanufactured parts. For example, officials told us that on January 8, 2013, 15 out of 1,037 purchase orders that day were coded under rebuilt.

[11] Many of USPS's delivery vehicles are custom-built to USPS specifications.

[12] The U.S. International Trade Commission reported similar findings, stating that numerous factors affect the demand of remanufactured vehicle parts, including the age of the U.S. motor vehicle fleet, the cost and availability of substitute products, component reliability, vehicle warranties, and the degree of consumer's cost consciousness. U.S. International Trade Commission. Remanufactured Goods: An Overview of the U.S. and Global Industries, Markets, and Trade, Investigation No. 332-525, USITC Publication 4356 (October 2012).

[13] When a vehicle is repaired, the old or failed part that was replaced is called a "core." A core charge is sometimes added to the cost of a replacement part and can be refunded to the customer when the failed part is sent back to the remanufacturer to begin the remanufacturing process.

INDEX

D

E

F

G

H

I

J

K

L

M

N

O

P

Q

R

T

U

V

W

Y